P9-CEL-691

Coroner's Journal

Coroner's Journal

STALKING DEATH IN LOUISIANA

Louis Cataldie, M.D.

Foreword by Patricia Cornwell

G. P. PUTNAM'S SONS

New York

ᴥP

G. P. PUTNAM'S SONS
Publishers Since 1838
Published by the Penguin Group
Penguin Group (USA) Inc., 375 Hudson Street, New York, New York 10014,
USA • Penguin Group (Canada), 90 Eglinton Avenue East, Suite 700, Toronto,
Ontario M4P 2Y3, Canada (a division of Pearson Penguin Canada Inc.) • Penguin
Books Ltd, 80 Strand, London WC2R 0RL, England • Penguin Ireland,
25 St Stephen's Green, Dublin 2, Ireland (a division of Penguin Books Ltd) •
Penguin Group (Australia), 250 Camberwell Road, Camberwell, Victoria 3124,
Australia (a division of Pearson Australia Group Pty Ltd) • Penguin Books India Pvt Ltd,
11 Community Centre, Panchsheel Park, New Delhi–110 017, India • Penguin
Group (NZ), Cnr Airborne and Rosedale Roads, Albany, Auckland 1310, New
Zealand (a division of Pearson New Zealand Ltd) • Penguin Books (South Africa)
(Pty) Ltd, 24 Sturdee Avenue, Rosebank, Johannesburg 2196, South Africa

Penguin Books Ltd, Registered Offices:
80 Strand, London WC2R 0RL, England

Library of Congress Cataloging-in-Publication Data

Cataldie, Louis.
Coroner's journal : stalking death in Louisiana / Louis Cataldie.
p.   cm.
ISBN 0-399-15282-2
1. Cataldie, Louis.   2. Coroners—Louisiana—Baton Rouge—Biography.
3. Medical examiners (Law)—Louisiana—Baton Rouge—Biography.   4. Forensic
pathology—Louisiana—Baton Rouge—Case studies.   I. Title.
RA1025.C38A3   2006   2005053474
614'.092—dc22
[B]

Printed in the United States of America
1   3   5   7   9   10   8   6   4   2

BOOK DESIGN BY AMANDA DEWEY

While the author has made every effort to provide accurate telephone numbers and Internet addresses at the time of publication, neither the publisher nor the author assumes any responsibility for errors, or for changes that occur after publication. Further, the publisher does not have any control over and does not assume any responsibility for author or third-party websites or their content.

The events described in this book are real. In some cases, names and/or identifying characteristics have been changed to protect people's privacy.

To my son, Michael

Contents

Foreword by Patricia Cornwell ix

Editor's Note xiii

ONE
Katrina *1*

TWO
Coroner *21*

THREE
Forensics 101 *41*

FOUR
What Are the Odds? *69*

FIVE

Too Young to Die   *91*

SIX

Head Cases   *117*

SEVEN

Final Exit   *143*

EIGHT

Thou Shalt Not Kill   *169*

NINE

Headhunter   *199*

TEN

Monster on the Loose   *209*

ELEVEN

Unsolved Mysteries   *245*

TWELVE

In the Sights of a Sniper   *255*

THIRTEEN

A Killer Strikes Again   *271*

FOURTEEN

To Catch a Killer   *295*

FIFTEEN

Conclusion   *323*

Acknowledgments   *333*

FOREWORD

I first met Louis Cataldie in the summer of 2002. While doing research at Harvard University for the book I was writing about the infamous serial murderer Jack the Ripper, I came across a newspaper article about similarly vicious multiple murders then occurring in the Baton Rouge area of Louisiana. Already DNA had linked the cases of three women who had been raped and murdered. In time, more victims would be discovered.

The Baton Rouge case—now solved, the killer on death row—would prove to be one of the most difficult and disturbing such cases we had seen in a long while. I suggested to ABC that we do a story on it for *Primetime Live*, hoping that the most important person in the story—East Baton Rouge Parish coroner Louis Cataldie—would agree to tell the truth about what was really happening in his small, violent parish. At the outset it was apparent that scarcely anyone wanted to talk about the murders, including the Baton Rouge police. Not even in the investigation of Princess Diana's death had I encountered such a hostile news blackout. The Baton Rouge authorities seemed to have no interest in disseminating accurate and helpful information about their latest serial killer (they've had more than one), and I began to fear that Dr. Cataldie wouldn't be any different. But he was. He was open. He didn't mince words. He was boldly honest.

In the course of doing that story, I saw firsthand how Louis made himself accessible to the devastated family members and friends of the victims, and how he became for them the family doctor they never—not even in their darkest nightmares—would have imagined they might someday need. Throughout a tragedy I began to think would never end, Louis consoled the bereaved and faithfully tended to his dead patients in a morgue fashioned from a trailer and equipped with hand-me-downs from funeral homes and restaurants. All the while having to work with officials who seemed hell-bent on gagging him, even running him out of office.

Three years later, two hurricanes—Katrina and Rita—have ripped through the corner of the world Louis loves so much. The scale of this tragedy—so many victims and such utter devastation for the survivors—would defeat most of us. But not Louis Cataldie. Sometimes sleeping no more than three hours at a time for days, he oversees the identification of the bodies of victims, counsels the suffering survivors, and copes with politicians and bureaucrats who get in his way.

In his native Louisiana, Louis Cataldie is a hero. He is, I think, nothing less than a national hero.

This book is the chronicle of a life spent doing work most of us would find depressing and grim. It is Louis's account of his passion for his agenda: dignity for the dead. It is told with both compassion and color, in a manner that is sometimes irreverent and never swollen with self-importance, by one of the most honorable people I have ever met.

—*Patricia Cornwell*

EDITOR'S NOTE

On August 29, 2005, weeks before *Coroner's Journal* was scheduled to go to press, Hurricane Katrina struck Louisiana and the Gulf Coast of Mississippi. Louis Cataldie was asked to assist in the massive evacuation of patients from Louisiana's State Emergency Operations Center at the New Orleans Superdome, an effort coordinated by Dr. Jimmy Guidry, the state health officer for Louisiana. In addition, Dr. Cataldie helped monitor the setup of field hospitals in and around Baton Rouge, gave medical attention to the injured, and began the arduous process of overseeing that every Katrina-related death was properly investigated. Then Hurricane Rita hit. With a medical system in chaos, and with more than 1,000 deaths associated with the two tragedies, the state appointed him Louisiana Medical Examiner. As of this writing, Dr. Cataldie is heading up the makeshift morgue in St. Gabriel, outside New Orleans, where the processes of identifying the victims and assessing cause of death are expected to take months.

—October 2005

Coroner's Journal

ONE

Katrina

"LET THE DEAD TEACH THE LIVING"

St. Gabriel, Louisiana—I am exhausted. I haven't had the luxury to reflect, or even to think—I'm simply on autopilot. My clothes stink and my head hurts, I haven't slept for more than three hours at a time for weeks; haven't seen my wife and my son but twice, and then for only a few precious minutes. My life seems so remote and far away.

It's been a month since Hurricane Katrina slammed into the Gulf Coast and tore a deadly path through the state I call home. I had always feared this day would come, as had anybody who lives around New Orleans, a city in a bowl below sea level.

"What if a cat-five hit N.O.?" De and I talked about it

many times. "You don't want to know," was usually my answer. Actually, we had been preparing and training for such an event since 2004; that ongoing planning, which we had labeled "Hurricane Pam," was interrupted by the real thing. We implemented that plan, but the component that dealt with the dead was simply "Call in DMORT." Little did we know that DMORT, the Disaster Mortuary Operational Recovery Team, was not prepared for such a catastrophe either. Over the next several months, I would become acutely aware of the limitations of the DMORT system. It was a tough lesson for us both.

And then, on August 29, it happened. Katrina slammed into the coast with 140-mile-per-hour winds, leaving a path of ruin in her wake. A month later, Mother Nature demonstrated her supreme indifference: on September 24, Hurricane Rita crashed into the Texas/Louisiana coast, deluging the area and pushing water back over the levees around New Orleans, which was still 80 percent covered from Katrina. Though Rita did not take the same deadly toll, the damage was incalculable.

I thought I'd seen everything. But now, as I look here at the remains of the dead and the empty eyes of their survivors, I feel utterly helpless. In the face of the annihilation of whole towns and neighborhoods, I am as lost and confused in this carnage as anyone. But then I ask myself, *Who else will account for the dead?*

I am still in the thick of it—more than 900 deaths in Louisiana alone, and I cannot see quitting anytime soon. In all, the storm has already killed 1,130 people in five states.

I take a reconnaissance helicopter flight over Cameron Parish, a coastal area near the Texas border that, since Rita, does not exist anymore. There is no town of Cameron. The only building left intact is the courthouse. Faint skeletons of what once were homes rise like matchsticks from the water, in a landscape that is desolate and eerie and that stretches as far as the eye can see.

The first person in charge I meet is Army Lieutenant General Russel Honoré, a fellow native appointed to clean up the Katrina mess and bring order to chaos. "Hi, we need tetanus," he tells me—that is about the extent of the conversation. The general, who said Cameron Parish was the worst he had seen in his survey, would later use a single military term to describe it: *destroyed*.

The only doctor who lives in the parish is Richard Sanders, a coroner and physician. He's a good guy, and he looks like hell—sweaty, tired, haggard, and going at Mach 2. In addition to no meds, no home, and no office—all just *gone*—he has more than a hundred open or floating coffins that had emerged from the ground because of rising floodwaters. We are making a plan to retrieve the coffins when we get news that a shrimp boat is coming in with sick folks on board.

No meds. No vaccine for hepatitis. In New Orleans there were dead bodies floating down the street. And now we have sick people coming our way. Our best defense is raw dedication—*you do the best you can.*

I meet an elderly couple whose home was carried about ten miles from its original address; they are on their way to the National Wildlife Refuge, which has become the Emergency Operations Center, to get a tetanus shot. I had just left there, and have to give them the bad news: The Center is out of shots.

Their son gives me a ride. He has lost his home and his shop. He points to a slab where his office was. The hospital, too, is destroyed. He doesn't know how many people will rebuild, and he scoffs a little as we drive by a "hurricane-proof house" that has been decimated. It was featured in *National Geographic*, he tells me—"But that was years ago." Rita caused no deaths here because people evacuated—in part because they remembered 1957 and Hurricane Audrey. I was nine years old at the time, and I remember that bitch, too. Katrina could have been her twin sister.

Most of the power lines are lying in the mud, and it will likely take months, if not years, to fix them. "We had to cut some wire just to get down the road—I think that may make them mad," the son tells me matter-of-factly. He strikes me as the kind of person who does what has to be done. He points yet again to a slab where relatives used to live, and stares there a little too long, giving himself away. I pretend not to notice the small crack in his veneer.

When we arrive at the courthouse, it is surrounded by military vehicles. The court clerk is there gathering records—he protected many of them before the surge. A surge is the deadliest part of the hurricane. It's basically a big dome, made by winds from the eye of the storm, that comes ashore as a wall of water. The clerk offers his office as a clinic if we need it. Inside, it's wet and already moldy, but it is standing and that is testament to the wills of these people. Some of the metal buildings are sheared off at about twenty feet above the ground. The surge must have been of tsunami proportions.

When I get back to the EOC, I talk to the doc and assure him we will get supplies. I leave with a handwritten list of medications he needs. There is no electricity. Stacks of plastic bottles filled with precious water sit in the sun. Oddly, they remind me of sandbags. The water is tepid if not downright hot, but you are glad to guzzle it down because it is clean. The heat and humidity rob you of your fluids in the merciless sun down here—even in September.

Dr. Sanders needs those meds now, and part of the reason I came here is so that he could look me in the eye and know who I am and have me as a direct contact. I put a lot of faith in seeing a man one-on-one. That's why I respect Stewart Simonson, assistant secretary for Public Health Emergency Preparedness at the Department of Health and Human Services, who met me face-to-face and has stayed true to his word. Imagine, someone from D.C. being a straight guy. He has delivered every time—not so with some of these other cover-your-ass feds. It's disgusting, and disappointing.

I made the decision to have chaplains go into the field with the recovery teams and say a prayer of thanksgiving for each person recovered. Incredibly, somebody from the ACLU had the audacity to take issue with my mixing church and state. My intent was to assure the dignity of the deceased. *No good deed goes unpunished, especially here.*

For many politicians, the point in coming down here isn't to lend a hand or to see for themselves how bad things are—it's all about the photo op! Can you imagine? Who in the hell is interested in token souvenirs when we have dead and dying?

In our makeshift morgue here, about seventy miles west of New Orleans, one of the forensic pathologists has hung a sign that reads: *Mortui vivis praecipant.* It means, "Let the dead teach the living."

RESCUE ME

During the rescue phase immediately after the first storm, I was transporting a little old lady who was crying because she didn't know where her grandchildren were. The building she lived in was damaged and the kids were missing. I asked her why they stayed during the storm; she told me there had been so many false alarms before. . . .

I tried to calm a young EMT who had promised an elderly woman he would go back to retrieve her when he had the right equipment to extract her. Then he lost her and could not get back, and he assumed she drowned. I fear for him: he cannot rejoice in the lives he saved, because he is stuck on the one he may have lost. We in the medical business tend to focus on the losses and not celebrate the wins. . . .

It's unspeakably sad to walk into a hospital and see the evidence of trauma and step around and even over dead bodies that lie in the hallways and everywhere else. No lights. No water. No ventilation. Then you have to worry about lead poisoning (as in, getting shot), so I carry my .40-caliber Glock as I wade through the sewer and chemical hazmat of the flood. Most of the lead hazard was eliminated by the police. . . .

I opened a hospital door and two big dogs jumped out at me. Thank God they were friendly—scared the crap out of me. I tried to catch them, but they ran off. Some dog packs have taken to hunting the only food around. The smells are horrible: death, decomposition, sewage, toxic chemicals, and black mold. I got a lungful of something bad last week during a door-to-door search in an apartment complex for the elderly. I should have been more careful with wearing a mask, but I got tired and careless; it took two days to breathe right. We had to put one guy in a hyberbolic chamber. Hepatitis A is a big worry. When I got my shots, I made the mistake of telling the nurse I was a doc; I think maybe she'd been dumped by a doctor, because she practically did a bone biopsy on me when she gave me my needle. Hurt like hell, but it's better than yellow eyes.

Hurricane Rita complicated things for everybody. Just as we were pumping dry from Katrina, Rita came and reopened the wound—gangrenous at that—so we will go back into the sludge and try to get our dead and help our state heal and re-unite our people with their loved ones.

As Rita approached, I had twenty-eight trucks tied together at the forty-acre morgue site—they were full of human re-mains. It had been suggested that they be moved, but I elected to keep them here rather than have a macabre caravan of dead bodies go north.

STRESS TEST

Some do-gooder chaplain invaded my personal space one evening. I was taking a few minutes to revive myself, enjoying a sloppy joe—army rations, in fact, an MRE, meal ready to eat. Quite fitting, now that I think about it, because this is as bad as any battle scene. Anyway, this chaplain comes up and sits a little too close for me to enjoy my slop. What bugs me is he's so clean—clean shirt, clean pants, clean fingernails, his hair neatly combed. He sits down right next to me. I'm on a bench, and trying to go blank for a few minutes, just to get the images out of my head for a while. He interrupts me.

"How are you holding up?" he want to know.

He gives me that TV evangelical smile—I wonder if he practices it in front of the mirror—and I am instantly irritated by his very existence.

Cool it, Lou, I tell myself. *This is exactly what the guy wants.*

Against my every natural urge, I'm nice, because if I go off, I'm going to *really* go off, and this guy would misinterpret it as affirmation of his intuitive worth. You know, the kind of guy who goes back and tells his buddies how much stress I was under and how he was blessed to be there for me. He'd be wrong. While he mumbles his rhetoric, I wonder how the hell he's so dry out here in the heat and humidity. You know, I've met lots of chaplains out here, and some are here for the right reasons and some are here for themselves. This guy was not the real thing, so I brushed him off. *Get the fuck away from me.*

Give me the real guy who has reddened eyelids and no smile and smells of death because that's what he's been around all day and he's got the look and his face is greasy and his sunburnt pores are clogged because he dares not wipe his face because his hands have been in gloves and he may have touched places that harbor disease from the wet sewage. He knows the deal, and he says, "Let me know if you need to talk." He pauses and adds, "I do." Give me that guy any day. He's the one out in the shit with me—he knows.

This day started off at the morgue, the biggest in the world, I am told, then off to the Joint Field Office (JFO), with all the politics there, then a visit to a proposed cemetery site, then a review of public record policy, then helping local leaders in Cameron Parish, then debriefing on same, then back to the morgue and more problems re Katrina versus Rita damage, then conversations with various state and U.S. senators regarding missing persons, then prepping for an interview with the press. It's been a long one.

PRESS HERE

I repeatedly refused to speculate on the body count, nor can I understand the media's gruesome obsession with it. It is almost always the first question in every interview, and in some ways, the least relevant. I thought I would lose it with a reporter who had the temerity to ask if the body count would match that of the World Trade Center terrorist attacks. *Who thinks like that?*

I don't lie to the press. I strive to be polite, but I am direct. And I am amazed sometimes that they let me in front of a camera. This is how one interview with a correspondent from the Australian Broadcasting Corporation ended:

REPORTER: There's been talk of upwards of 10,000 bodies. I understand you've set up a morgue for 5,000 in St. Gabriel, can you tell me how big you believe your operation is going to be?

LC: Number one, we're worried about the individual and not about the number, and if there are 500 individuals, we will treat them as individuals. If there are 1,000 individuals, we will treat them as individuals. I don't know how many individuals we're going to find, I don't know how many family members we're going to find. We certainly have prayed for the dead and we're preparing for the worst, and if it's 1,000, or if it's five, we're going to do the right thing, so I don't want to speculate on numbers. As far as I'm concerned, until a body comes to my morgue I will not count that individual as a victim, because I just don't know. So I'm not going to speculate or guess on numbers, but I will tell you, yeah, I'm prepared for 5,000, but I certainly hope that doesn't happen.

The press has been both a help and a hindrance. There have been many questions about crime down here. My crew and I have been inundated with false reports. I heard stories of euthanasia, women being raped, people murdered. Police were supposedly in shoot-outs inside shelters after the storm.

When the smoke cleared and the water receded, four deaths by gunshot wounds were initially confirmed in New Orleans after Katrina. To put that in perspective, four's a typical week in the Big Easy. So while the press deserves credit for arriving at the disaster scene before the federal government did, I nonetheless fault the national media for not following up more aggressively on any of these rumors. Too often they simply accepted rumor as fact. As I told one reporter, "It's not consistent with the highest standards of journalism." In total, ten corpses were recovered from the Louisiana Superdome—contrary to urban legend, not one was a homicide.

DMORT

The tiny town of St. Gabriel, where we are based, was named after the archangel who in the Bible bears good news to people. Gabriel first appears in the Hebrew Bible in the book of Daniel. He is the messenger of God. He has high standing in the Christian faith for having announced the coming of Jesus; some Christians believe his horn will signal the second coming of Christ. In Islam, he is credited with revealing the Koran to Muhammad. He spans many faiths. What better guardian for the Disaster Portable Morgue Unit (DPMU)?

The town is now temporary headquarters for the Louisiana branch of the Disaster Mortuary Operational Recovery Team (DMORT), a mobile disaster squad made up of medical examiners, pathologists, and funeral directors from all over. (There is another in Gulfport, Mississippi.) DMORT was set up so that experts in nearby cities, acting on behalf of the federal government, could quickly relieve overwhelmed local resources in a disaster.

When a body comes through this morgue, it is decontaminated and assigned an ID number, along with an escort who will follow the body through this laborious but necessary process. A forensic pathologist will thoroughly examine the victim for any signs of foul play. To the husbands, wives, brothers, sisters, and children of the deceased, I can assure you: No one will get away with anything on my watch.

At another station, forensic dental experts take photographs and X-rays of the victim's teeth. DNA samples are also taken.

Bodies and personal effects are stored in a refrigeration truck until final ID and family notification are made.

The causes of death vary. Many are drownings. We also see chronic illnesses, predisposing illness, ventilator-dependent patients who died when electricity was lost, not to mention acute myocardial infarctions related to the stress and the trauma. Each body that comes in is a worst-case scenario for the victim, the family, and the coroner: there's nothing worse than somebody's dying like this.

Families of missing persons call into the Family Assistance Center (FAC) on a daily basis. We've had over 16,000 calls and more than 10,000 missing persons entered into the database. *Where are they?* Some are in the morgue, unidentified. Some have washed into the river and the gulf, some may be in the marshes, some may be in their flooded homes awaiting their family's arrival. What a horribly traumatic event, to go back to a flooded home and find the decomposing human remains of a loved one there.

They find their bodies in the toxic sludge, under furniture that floated about and under the drywall ceilings that crashed down on top of them. *Why weren't they found during the searches? For lots of reasons, there is no one answer.* I am relegated by those reasons to the position of trying to control the damage those families experience. I have been in the field talking to them. It sucks!

There have been problems with the DMORT database. It wasn't set up to handle missing persons and fatalities scattered over an area the size of Great Britain. I'm told it's held together with high-tech Band-Aids. I'm not even sure what that means.

The FAC members collect all the information on an eight-page form. They list the surgeries, prosthetic devices, unique characteristics, tattoos, and personal effects of the victims. Visual identification is a rarity, owing to the condition of the bodies. There will be no open-casket funerals. I need dental records, but many were destroyed in the flood. We send dental teams out into the flooded offices to try to salvage any records. We take photos of the victims' mouths and show them to family members for identification. Forensic dentists match pictures of smiling victims to their dental photos and X-rays here at the morgue. I always have one of the Victim Relief chaplains here for support when we do this. Such reality without compassion is indeed brutality.

I see those victims smiling back at me in those family photos. I hear them say, "Find me and send me home."

Fingerprints are great tools if you can get them, but the sludge and the water and the time and the animals have not been kind to these bodies. In short, some don't have fingers for us to print.

We have the same problem with medical records. Many doctors' offices and hospitals are flooded and out of commission. We have taken total body X-rays on all the victims. I hold a chest X-ray up to the light and look at the wire pattern in the sternum of an elderly man found at a certain address. He's had coronary bypass surgery. I have a lead. If I had his old chest X-ray, I could match that wire pattern and make a solid identification. Alas, the X-ray has suffered the same fate as the dental X-rays. We will not give up, we will continue the search. Every day I sit in the "IR" trailer with DMORT investigators, radiologists, forensic pathologists, and forensic dentists. Every day we try to reunite the deceased with their families. DMORT isn't some federal bureaucracy, DMORT is people—dedicated people.

There are a lot of dead folks here, and a lot of families wanting to know whether their missing loved ones are among the deceased. One of the most crucial and difficult parts of the job is to notify the families of victims and deal with their immense grief. Compunding the grief are the facts that families aren't able to view the bodies quickly, and that the bodies aren't released to the families fast enough. It's excruciating for relatives, but there is just no way I can speed the meticulous work of identifying the dead and preserving evidence of possible crimes.

As I told reporters on national television, "We won't take anybody out of the rotation or the process. To take one person out would corrupt the process." And so when folks say, "Can we speed up the process?" the answer is, unfortunately, no.

STAYING SANE

People always ask me how I stay sane doing what I do. I'm not sure what "sane" means in this profession, but it does take an emotional toll. I sometimes wonder whether maybe I've brushed shoulders so often with Charon, that mythological ferryman of the dead, that it's caused me to question the very nature of humankind. There's a fine line between healthy skepticism and jaundiced cynicism.

If there's one thing that's helped me walk that line, it's the fact that I've kept a personal journal, starting in the coroner's office of East Baton Rouge Parish in 1993. I wrote whenever it struck me, jotting notes in the field and later sketching images on a small pad as I remembered them. These were not sketches of any forensic value, but nuances that stayed with me for some reason. I wrote in my journal mostly at night, especially after a particularly troubling autopsy or a visit to an unsettling crime scene, when neither a soft pillow nor the comfort of DeAnn, my loving wife and partner, could induce slumber. The journal is about how the lives—and untimely deaths—of the people I investigated crossed my path, and how I tried to bring order and integrity to the aftermath.

I've become quite aware of the importance of these private journals because, since Katrina, I've been unable to keep up the practice to the degree I'd become accustomed. I simply haven't had the time. Other than the occasional flurry of midnight e-mails (from which this chapter evolved), I've been without my usual coping mechanism to process these experiences privately. The burden then falls, even more than usual, to De and Michael, whom I've barely seen in weeks and weeks. I miss them more than I can say.

Michael, my youngest son, is another reason for keeping these journals. I want him to be able to understand what was going on in his family during his formative years. How many eleven-year-olds go tracking across the shopping mall to decribe the "splatter pattern" of a leaky garbage can that has been moved across the floor? "Look, Dad, they stopped here for a second—see how the drops fell straight down? They started back up this way, and hey, look, over here they started speeding up. I guess they figured out they had a leak, huh?" Kids do tend to listen to our conversations, don't they? Michael once accused me of going to the movies without him—he found trace evidence of a popcorn hull on me. Actually, I was innocent: I had had popcorn at the office earlier. At any rate, I just want to give him some perspective as he wanders down life's road and takes his place in society. He may find it helpful later to reflect on how my years on the job affected him.

As coroner, I have served as the state's official witness to the worst that humanity has to offer. I've investigated some of the most despicable crimes and violent deaths imaginable, have held the limp bodies of innocent children. Nothing, not even the profound inhumanity of serial killers like Derrick Todd Lee and Lee Boyd Malvo and John Allen Muhammad, can compare to the sheer devastation of Katrina and Rita. Yet in many ways, it seems that everything I've done in my professional life has been training for these last terrible weeks. The stories that come later in this book trace my evolution not just as a doctor but as a human being. The people I met, the families I counseled, the mistakes I made, the crimes I helped solve—these were my training ground, I see now, for the biggest challenge of my life.

I didn't have a survivor's guide when I got into the coroner business. I certainly could have used one, though. I don't think one exists. Maybe this will help some of the other folks who get into this vocation and have to deal with and reconstruct the details of a person's final minutes or moments before death comes.

Be forewarned. This book is no *CSI: Cajun Country*. I'm not a polished movie star, we don't rely on special effects, and the stories of the people who are left holding the pieces can't fit neatly into a one-hour episode. It's about emotional survival on the battlefield of death investigation. Death will walk with us.

On the surface my job seems quite simple. I pursue the cause, time, and manner of death. My responsibilities are to the deceased, the survivors, and society at large. There are times when I feel it is the survivors who suffer the longest if not the most. The violent or unexpected death of a close friend or loved one can shake the foundations of a person's existence. Some of those left behind in the aftermath of a death may suffer from pathological grief disorders or post-traumatic stress disorder. They live with recurring nightmares and intrusive thoughts that can plunge them downward into an abyss of severe depression and hopelessness.

And as a coroner, I, too, have tasted from that deadly cup. At times, I've come away from crime scenes and autopsies with a tainted worldview: tainted by anger that I strive not to bring home; tainted by skepticism turned to cynicism; tainted by fear. That taint has been lessened in part by people like Caron Whitesides and Wanda Hebert, who have worked with me for years and shown me how to salvage some good from the bad.

There is a truth I accept on this matter: *If you don't deal with a horrible or traumatic event, it will deal with you. There is no escape.* In recent weeks I'm having trouble taking my own advice. I accept the reality of the situation, do my duty to the best of my ability, and deal with the emotional consequences.

NOT OVER YET

I have witnessed unbelievably selfless acts from fellow coroners, rescue workers, and medical teams from around the country. I have been inspired by the downright unflappable spirit of the people around Cameron and the natives of New Orleans—it makes me proud to be from Louisiana. But there is no solace when you look at the hurt in the eyes of the survivors. It is my obligation to do it right; 99 percent is a failing grade. The feds will go away, but I will not.

What went wrong? Why weren't the local, state, and federal governments prepared to cope with a hurricane or disaster of this magnitude? I'm sure politicians will appoint committees to investigate, as well they should, and the blame will fly faster than the mud in a political race down here. I sure as hell can't explain it—truth is, I'm not even going to try, at least not here. All I want to do here is to assure the survivors that dignity for the dead is my only agenda.

Finally, I ask readers to forgive such an incomplete accounting. There aren't words invented yet to describe the depth of despair—or courage—in this corner of the world I love so much. No one person can attempt to tell this story. And it is not over yet.

Gotta go.

—Lou Cataldie
September 29, 2005

TWO

Coroner

FIRST BLOOD

My first death investigation came on a gorgeous spring day in 1976. I was standing on the porch of a beautiful white turn-of-the-century house graced with tall windows and delicate Victorian touches. This house—across the alleyway from the courthouse, a cold brick structure that housed the sheriff's office and an un-airconditioned jail on the third floor—was the place where I worked. It had been converted into a medical clinic, and I was the only medical doctor in this rural outpost. I loved the feel and sound of walking on a wooden porch, especially this one—built out of cypress from the local sawmill. Though the interior had suffered from some modernizing attempts in the 1950s, the rooms were spacious and had twelve-foot ceilings. We had furnished the facility with medical equipment that was military salvage from a MASH unit—circa the Korean War by the looks of it—but it worked.

I had decided to follow my altruistic inclinations and become a general practitioner in a small rural town in a state I hold close to my heart. Looking back, that altruism was mixed with an element of ignorance on my part—my desire for a sort of Norman Rockwell life. At any rate, I took my new wife and son to the tiny north Louisiana town of Colfax, population 1,800. There was only one other physician in the entire parish. It was about a year before that altruism was beaten out of me. While I was there, though, I became gratis coroner of Grant Parish, as well as the jail physician; both positions came to me by default. The other doc had served his time in the barrel and had no desire to continue the post. That's called a clue! But who could blame him? There was no budget, and expenses, such as paying for an autopsy, had to be approved by the police jury.

That morning, a breathless young man burst through the doors of a full waiting room while I was seeing patients. I ran out to the man, who kept yelling something to the effect that *"the Colonel need to come quick."* It would not be the last time that I heard myself, the coroner, being referred to as "Colonel." I rushed out of the examining room and followed the man onto the porch. I was expecting to be greeted by a medical emergency. Instead, what I found was an obviously dead body in the bed of a road-worn pickup truck.

Of course, all my patients had rushed out behind me to peer at the corpse, too. The two guys in the cab of the old truck jumped out and explained what had happened:

"Zack here got hit by a piece of pulpwood what fell off the truck," said the first man. "It landed right smack dab on his head! That's what happened, Colonel. We got the pulpwood log right here for you to see, uh-huh."

The fact that Zack was dead was apparent to all involved, and by now the whole town was involved, including the sheriff's office. Zack's skull was caved in, and the alleged cause of his demise—a rather hefty log that looked to be about eight inches in diameter and must have weighed several hundred pounds—was there as well.

The witness continued: "We was way out in the woods and we didn't know what to do except to bring him here to you, Colonel."

"I'm not a colonel. I'm the *coroner.*"

"Okay, Colonel, sir. What we supposed to do now?"

Everybody around the truck—a sizable crowd had gathered—nodded their heads in understanding and offered opinions.

"Pulpwooding is dangerous business all right."

"Look, there's the log right there."

"Kilt him dead."

I jumped up into the bed of the rusted green truck and examined Zack. He wore a faded denim shirt and jeans and was lying on his back staring up at the sky. He wore scuffed and battered tan work boots. The right side of his head was literally smashed in. Abrasions on his torso were compatible with the story presented by his coworkers.

The truck bed was wooden and weathered. Zack had been bouncing about in the back of the truck during the heated journey from the piney woods to my clinic. He needed a shave and had brown hair. I judged him to be in his early forties, but he was actually in his mid-thirties. Life is rough for a pulpwooder. The sky was clear and blue. It was such a peaceful day in Colfax. Death comes to Mayberry.

There was no need for an autopsy. The sheriff's office did their investigation and it was over. There wasn't really much to do. The cause of death was head trauma and the manner of death was accidental.

Twenty-five years later, I still have a 3-D color image of that moment in my mind. I see myself in my white doctor's coat, crouched over poor ol' Zack, trying to see if the story fit the evidence.

I was hooked and I didn't even know it.

SOUTH CENTRAL LA

Before I get started, I feel obligated to give you a little information about me, the responsibilities of my office, and some of the things that make Louisiana unlike any other place in the United States.

Two types of forensic death-investigative offices or systems exist in this country: coroners and medical examiners. A medical examiner is a licensed medical doctor, often with a specialty in forensic pathology. A coroner, on the other hand, could be anyone who can get elected to the office. The coroner does not have to be a physician in some areas. In Louisiana we have three nonphysician coroners that I know of. If the office is not filled by an M.D. (usually because nobody wants it), a nonphysician may run for the office. But if an M.D. wants it, the M.D. gets it. I am not a medical snob, but having a doc as the coroner—a professional versed in the knowledge of medicine and human physiology—is a good thing.

Medical examiners are appointed and therefore do not have to face the political gauntlet. Some argue that this makes the ME system a better one, but there is a critical difference: the ME works at the pleasure of his or her employer. So in the end it all comes down to someone's integrity and their skills in the office. To complicate matters even more, some states have both coroner and medical examiner systems in place.

In East Baton Rouge Parish, an area of about 400 square miles, I investigate about 2,000 deaths a year—around 60 homicides, 30 suicides, and 150 traffic fatalities, most of which involve a person driving drunk or sleeping at the wheel.

Many cases are "routine"—if there is such a thing as routine in this context. The identity of the deceased is known and the family is easily contacted. In other deaths, critical information is missing. Identification may require dental records, DNA, or fingerprints. I call on anyone who can help me solve an open case, from forensic anthropologists, who can tell the sex and age by looking at bones, to forensic entomologists, scientists who study the life cycle of various insects that feed on corpses and can help determine time of death.

Like that of other coroners, my office is responsible for investigating the time, cause, and manner of death. The body is our evidence, and we preserve evidence—anything from a blood sample to a body part—for examination and analysis. At autopsy we take photographs and/or direct the police to do so. We provide testimony at depositions and in court. My staff and I meet with police, doctors, and lawyers to answer questions in criminal and civil cases. We train law enforcement, aid victims' groups, and inform the media about our role and function.

An act of the Legislative Council of the Territory of Orleans, in April 10, 1805, created the office of the coroner. The duties in 1805 were, simply stated, to deal with the dead and with any death investigation. In 1814, the coroner was given the power of arrest, and today the coroner remains a conservator of the peace. Coroners were charged with pauper burials, which also continues today. In 1846, the office was changed from a governor's appointment to an elected position, making it subject to the Louisiana election process. The coroner was not mandated to be a physician until 1879.

In Louisiana, though, the duties of the coroner are different from those required by law in the rest of the United States. In addition to our death investigation duties, we have mental health commitments and carnal knowledge (rape) duties. The evidence I find can bring comfort to a grieving mate or send a man to death row.

Under the mental health law in Louisiana, the coroner is responsible for involuntary commitments to mental institutions. A family member can request a pickup order, or OPC (Order of Protective Custody). If the coroner believes a pickup is warranted, he'll sign an OPC and the subject will be picked up by the police and brought to the office or to a mental institution.

Perhaps the most difficult part of the job is stipulated by the carnal knowledge law, which makes the coroner responsible for the examination of all rape cases. The Baton Rouge Women's Hospital does this through its rape crisis unit—unless no one is available, in which case I give the rape examination—and the victims include not only female but also homosexual-male rape victims.

The job is slightly more complicated than the state itself. Louisiana is divided into sixty-four parishes, which are the primary local government instead of counties, a legacy of colonial Louisiana, which was officially Roman Catholic under both France and Spain.

If you want to understand the context, it's best to view Louisiana as sort of a composite state. To the northwest is Shreveport, which, locally, is often considered to be a part of

Texas. To the northeast is Monroe, home to rednecks mostly. (I should know—I went to college there.)

My hometown, Alexandria, is in the center of the state—the ankle of the boot in a Louisiana map. It's a humdrum town that has never risen to its potential. But Alexandria, or, more accurately, the little town of Bunkie to the south, is (again, locally) considered to be the demarcation line between North and South. Welcome to the Deep South and the Deeper South.

Alexandria is also the northern point of the "French Triangle," with Lake Charles in the southwest and New Orleans in the far southeast. In between is Cajun country, inhabited by descendants of the state's earliest French settlers, people with names like Boudreaux, LeBlanc, and Robichaux. Folks in these small towns generally have a very strong sense of family, and many still speak Cajun French as their first language. It's a rich and colorful culture.

New Orleans, known by many as the Big Easy, considers itself to be a state in and of itself. It is a town where you can legally gamble, drink openly in the street, and buy a ninety-proof daiquiri at a drive-thru window.

Not quite so far to the southeast lies Baton Rouge, the state capital, where I live today and where I have investigated more than 300 homicides and thousands of other deaths. It is a green city with lots of stately oaks, beautiful antebellum homes, and river mansions. From the Mississippi River Bridge, you can see the needle point of the capitol building. Below, the mighty Mississippi cuts a wide swath between East and West Baton Rouge parishes. Baton Rouge is a little big town: little-town flavor with Southern manners and big-town problems. The population of the combined parishes is a little over 400,000 with over 200,000 people residing in the city proper. It holds the unchallenged title of having had the longest-running school desegregation case in the history of the United States—over forty-five years, and just recently resolved. Ironically, the white public school students are now in the minority. The city suffers from "white flight" and a shrinking tax base. It has two universities—LSU and the black college, Southern University.

We have several police departments: the Baton Rouge PD, the East Baton Rouge Parish Sheriff's Office, the city constable office, the LSU police, the Southern U police, and two other city police departments in the parish in Zachary and Baker. Most of the crime and homicides occur within the city of Baton Rouge, however.

And the city bears the brunt of poor planning: major traffic problems, eroding tax base, and high crime. As if that weren't enough, Baton Rouge is number two in the nation for new AIDS cases (per capita). Leadership, as always, is elusive. The last three secretaries of insurance have been sent to prison (Jim Brown served six months for lying to an FBI agent and is now out; Doug Green, his predecessor, is serving a twenty-five-year term for taking $2 million in illegal campaign contributions; and Sherman Bernard pleaded guilty to taking bribes and served thirty months). Governor Edwin Edwards is finishing out his ten-year sentence (for racketeering)—he was convicted of extorting money for river boat gambling licenses, and is appealing his conviction. The secretary of agriculture was under indictment initially for twenty-one counts of theft, money laundering, filing false public records, and bribery, but all were ultimately dismissed. The point being he was elected while under indictment. Louisiana politics are indeed "unique."

FAMILY MAN

Like many people, I have found that many of my choices in life were dictated by family and circumstance. My father was a first-generation Italian-American and Catholic. His father was from Sicily and his mother from mainland Italy.

My mother was neither Italian nor Catholic. She was not only a Baptist, but a hard-shell Baptist, who avoided the evils of music (except gospel), dancing, and the seven deadly sins. Every now and then she'd drag me off to a tent revival. She came up hard and poor in the northern part of the state on what could be described euphemistically as a farm. She was never really accepted into my father's family.

I, on the other hand, was totally accepted and embraced by the Italians *and* by the hillbillies. I was blessed with warm, caring grandmothers on both sides—my Italian "Mommee" and my hillbilly "Ma." I loved my maternal grandma very much. She was a really good person who used to read the King James Version of the Bible to me. I've even been known to join in the hymn singing with her up there on Breezy Hill at tent revivals.

At any rate, this was the backdrop, or battlefield, as the case may be, upon which my religious perspectives developed. Though I gradually drifted away from the Church, I think having a foot in several diverse belief systems allowed me to learn that diverse cultures can coexist. It was only later that I realized how valuable it had all been. Today, I'm still not into any organized religion, but I do believe there will be an ultimate accounting for the life I live.

Growing up, I found my career choices to be rather limited. The most obvious: follow in my father's footsteps, working in a café or a grocery store for the rest of my life. My grandfather had a gambling hall, with cops on the take per family legend. I was exposed to crime and the frailties of the police at an early age. I guess my other (unstated) career choice was to become a bookie in Alexandria.

Fortunately, Brother Cosmas of the Sacred Heart Order stepped in and changed my life forever. I was in Menard High School, a Catholic all-boys school run by the Brothers of the Sacred Heart. Brother Cosmas was a no-nonsense kind of guy who informed me one Monday that I would be taking a test the following Saturday. I still have a vivid image of him—a tall, gruff-looking guy, but with a heart of gold. He actually practiced what he preached in that cassock, and he stood for something. People feared him, yet I knew for a fact that he regularly fed a skinny old stray dog that came by the side of the handball court on the school grounds. The test turned out far better than I had expected: it led to a full academic scholarship to Northeast Louisiana State College. I graduated with honors and was accepted by two medical schools.

I thought this was the perfect time to get married. It wasn't. My first marriage was a complete mismatch, for both of us. I was immature and had no idea what to expect. It was a stupid thing to get married and then go off to medical school. I was never home, and even when I was, I wasn't. The bottom line is that it turned out to be a crappy deal for her, and I take responsibility for the failure. I didn't have another healthy marriage to model mine on, and that just compounded the problem. The marriage lasted eight months before we divorced.

When I entered LSU medical school, the Vietnam conflict was in full swing, and high school classmates started coming home in body bags. I fully anticipated going into the military during my internship at Lafayette Charity Hospital, but as was to become the norm in my life, fate intervened and the war ended.

So I got married. Again. We ended up in the picturesque north Louisiana town of Colfax. Hooker's Café served great home-cooked food. Everyone knew everyone. And the folks were nice, though the money and the hours were horrible. I was paid with fish, deer meat, fresh vegetables—which I respected, but I would have preferred cash. Most of the folks were poor but proud. The closest hospital was in Alexandria, thirty miles away.

I was in over my head. They needed a seasoned physician, not a tenderfoot just out of med school. I think I would have self-destructed had I stayed there any longer.

Over the next year or so I investigated several accidental deaths, one open-and-shut homicide, and two suicides. There were also a few natural deaths—no mysteries.

Being a GP in a rural town wasn't working out, for my wife or me. My wife was from Baton Rouge, and when an offer to work there as an emergency room physician came my way, it seemed awfully attractive. The year was 1976, and at the ripe age of twenty-eight, I escaped with her to the "big city."

"RED STICK"

In French, Baton Rouge means "red stick," and some say the place was named for the giant cypress tree where the settlement was located. Darker theories hold that it was named for the blood of an Indian massacre at the spot. Most folks just figure it was a skinning post. Take your pick. I joined an emergency physician's group there in late 1976.

The emergency room life was a frenzied, high-stakes test of wits and adrenaline, and I took to it right away. I was based at Baton Rouge General, one of the two hospitals in town. Though it is now in a bit of decline, I remember the floors were always clean and shined—so much so that I slipped one time when I was running to answer a code 14—cardiac arrest. Those were the "good old days," when there was no ER specialty and we covered emergencies in the whole hospital.

The place felt like home. The brave nurses I had the privilege to work with were always trying to make things better, despite rowdy drunks and violent PCP addicts. The action was quick and it was most rewarding to save someone. What kept me and others juiced was the element of suspense: What would be coming through the door next—an MVA (motor vehicle accident), a GSW (gunshot wound), or a stabbing? In retrospect, I think we were all adrenaline junkies. We not only liked to win, but we took defeat personally. Defeat, whether by natural or unnatural causes, was unacceptable. That attitude of perfectionism took its toll on all of us. There was no such animal as "critical incident debriefings" or Employee Assistance Programs in the late 1970s. We were on our own down there.

When we saved a life, there was jubilation. When we lost, we simply drifted away from the deceased and never really processed it from an emotional perspective. Some of us would drink over it later at the nearby Fleur de Lys Lounge. The ER also proved to be a good training ground for exposure to the criminal elements in the city. Those folks just tend to get beat up and shot— a lot. We also handled drownings, child abuse, accidents, motor vehicle crashes, and suicide attempts. During this dawning era of designer drugs, angel dust, or PCP, was plentiful, and drug overdoses and "bad trips" were also commonplace.

Unfortunately, my own propensity for "all-or-nothing" behavior resulted in a short time in therapy, to get my own head back on straight. It was a good thing. I have been free of addiction for twenty-six years now. In addition to being one of the founders of the Physician Health Plan for Louisiana, I am also the consulting medical director for the state's Office of Addictive Disorders. I hope my experience helped break the family curse.

Though my official title was Physician Executive in Charge of Medical Affairs for the Psychiatric Division, I stayed active in the actual practice of medicine.

One day I took a break from rounds for a cup of coffee in the doctors' lounge of the Baton Rouge General Hospital, when I was approached by a doctor with the unusual name of Hypolite Landry. He was the coroner for East Baton Rouge Parish, and he needed a deputy coroner. During our talk, he reminded me that in Louisiana, the coroner is responsible not only for death investigations but also for mental health commitments. I liked the mental health part, or so I thought at the time, and took the job as deputy coroner right there on the spot.

My expectations burst one morning in June 1993, at about one A.M., a few days after I had impulsively taken the job. The answering service called me to report a death. I picked up on the second or third ring—years of being on call made me a light sleeper. At first I thought it was the hospital calling me about one of my patients. No—this was the coroner's office service. I was informed that no one else was answering their pages, so it fell to me. I ran out the door completely unprepared for what I would see next.

The deceased was a gunshot victim out on Airline Highway, which runs north–south through Baton Rouge and continues another hundred miles down to New Orleans. My directions were simply that it was near the fairgrounds. I was almost out of the parish by the time I came upon the flashing blue lights. When I arrived at the scene I had to identify myself. The "gate-keeper" looked somewhat puzzled and responded as such:

"What got you out, Doc? Nobody else around? It's down there. Which funeral home you want me to call?"

Now it was my turn to be puzzled. I had no idea about which funeral home to call. So I was flagged through to where the body lay. Once I got within range of the body, a detective began waving his arms and yelling for me to stop. Again I was puzzled. When I got out of the car he told me which path to take to the body.

He was trying to make sure I didn't bumble into the scene

and mess up the evidence. That feeling of being ill prepared was rapidly turning into embarrassment, and I started to feel really dumb. *You're not in Kansas anymore, Dorothy—nor are you in Colfax, Lou!* I was getting my first lesson in crime-scene etiquette. It would be the first of many.

I introduced myself to the detective and then we just sort of looked at each other. Obviously there was some expectation here.

He broke the silence. "Hell, just a hundred more yards and he'd be in the next parish and I'd be having breakfast at the Waffle House. I guess it's too late to drag him across the line."

I stared at him. He pointed to the corpse.

That was a joke, dumb-ass.

"Looks like a gunshot to the head. We pretty much know what happened. Shot over the P-word," he said, referring to the vernacular for the female-gender-specific body part. It was an accepted truth in this business: men tend to shoot other men over women. It's an ego issue, often exacerbated by alcohol, and it goes by a lot of names—possessiveness, insecurity, lust, love, pride, false pride, male stupidity, power trip.

"You want me to dispatch a funeral home?" the detective asked.

I was essentially lost. I decided to just come clean with the detective. "Look, I'm new out here. What do you need me to do?"

He didn't seem that surprised. I was like a calf looking at a new gate—*what now?*

The detective turned out to be a good guy and told me the coroner usually examines the body, turns it over for them to take some pictures, gets the ID off the deceased, and calls the funeral home. Then the next day the office calls the police and tells them when the autopsy is going to be done. That's it. The expectations on the part of the police of the coroner were limited, to say the least.

Even to my naïve way of thinking, that didn't seem to be what coronering was all about. But I went through the motions and called the funeral home—white funeral home for white victims, black funeral home for black victims. The funeral home would take the body to the Charity Hospital for autopsy. I later found out that I was to call the pathologist at Charity to get a time. I was not to call him at night—rather strange, but hey, that's the gig, Jack!

The victim had a gunshot wound to the right side of his head. There was no exit. That's about the extent of what I had to offer.

The police would later chalk the motive of this murder up to a woman. Police, who knew the deceased, Jerry David Dixon, twenty-two, picked up the suspect, Robert Lee Duke, not long after. Duke said he shot Dixon in the head after Dixon confronted Duke and his girlfriend, Ann Marie Tuccio, in a pickup truck on Airline Highway. Dixon had a former relationship with Tuccio, police said.

Duke was charged with second-degree murder in the slaying, but a jury convicted him in 1994 on the lesser charge of manslaughter. He is currently awaiting his appeal.

An interesting side note that came out during the trial was that Tuccio, in her testimony, reported having nine separate personalities. These ranged from that of a five-year-old girl to that of a topless dancer. After Duke's conviction, she was charged with being an accessory after the fact. She pleaded not guilty.

That was my first murder in Baton Rouge, but I knew that I would be better prepared next time around. There were many next times, for Baton Rouge would live up to its gory name.

THREE

Forensics 101

DO THE RIGHT THING

I felt I had a handle on the mental health duties of the job, but the realm of death investigation was intimidating. I was still smarting from the lesson in humility I'd gotten out on Airline Highway. To correct that deficiency, I attended every training seminar I could find and afford. I read forensics textbooks voraciously and went to just about every homicide crime scene in Baton Rouge.

One of the most helpful things I learned was crime-scene etiquette. A pearl of wisdom came from an older homicide detective, now retired. It had been passed down to him years before by his senior partner. "When you go through a crime scene for the first time, keep your hands in your pockets." He laughed when he said it, but it was no joke. "That does two things: it shows you're not stealing anything or planting evidence, and it shows you're not contaminating the scene."

That makes sense.

I sought out experts in every forensics field, and I was a sponge for their knowledge and experience. The crime-scene officers taught me how to collect evidence and examine latent fingerprints from corpses. We used alternative light sources to look for trace fibers and hairs as well as body fluids. I also learned about forensic photography and how to take casts of shoeprints.

Dr. C. Lamar Meek, a forensic entomologist and renowned mosquito expert who happened to teach at LSU, coached me on the valuable information insects can tell us, and the best ways to collect them. I was amazed to learn that maggots from a corpse can be put into a blender and made into a slurry— pathologists call it "the maggot milkshake"—that can tell us if the deceased was a drug user, and what drug was used. Some flies, he went on, prefer the dark recesses of an unlit room, while others prefer the sunny outdoors. So a corpse recovered indoors with the larvae that live in outdoor locales indicates that someone moved the body.

An internationally regarded forensic anthropologist at LSU, Mary Manheim, also known as "The Bone Lady," taught me how skeletonized remains can tell us who they are from dental X-rays. Scrutiny of bones may also reveal knife nicks and indicate cause of death. Pattern impressions from skull fractures can tell us a lot about the murder weapon.

The crime lab tutored me about DNA and sex crime investigation, and basic forensic techniques in hair and fingernail analysis, as well as ballistics. I have been with them in the field on several occasions; most recently during the hunt for a serial killer.

Arson investigators gave me a priceless education about hot spots and how to reconstruct a fire—how to read "alligator wood" and other signs of pattern burning. Important to know, as murderers sometimes use fires to conceal their crimes. More important, I learned how we could reveal those attempts. That skill set came in handy several times.

When I became coroner, there was no such thing as a seamless chain of custody for the bodies. They were simply picked up by whatever funeral home we could get to come to the scene. Much to my chagrin, the funeral home even had to supply the body bag at times. It was essentially a community service on their part, and they ran the risk of never getting paid for their services. When I was a deputy coroner, I went alone to crime scenes, but afterward I began taking someone with me.

Unbelievable as it might seem, the state's capital city did not have a coroner's morgue. Autopsies were done at a local hospital. There was no real security there. What's more, the pathologists were employees of the hospitals, which opened up conflict-of-interest situations. Eventually, we established a seamless chain of custody, and my staff began picking up the deceased in our own coroner vans. We built our own morgue in the form of a Medical Disaster Unit (a trailer). We have our own forensic pathologist and a new forensics facility is near completion.

As chief deputy coroner, I was eventually persuaded to assume the position of coroner when the old coroner retired. I won a special election and was unopposed in the following regular election, and here I am right now. I've been here ten years. But . . . why do it? I've pondered this a great deal. In the end I always come up with the same answer: *Because it's a job that's got to be done, and done right.*

RATS

In my three years as deputy coroner, I learned a quick lesson in crime-solving: criminals talk. Before I got into the coroner business I never really gave this a whole lot of thought. This basic concept was brought home to me just as I was becoming a full-fledged coroner in 1996, and it has proven true over time.

Shelia was admitted to Earl K. Long Hospital (EKL) and died there. For whatever reason, her physician did not deem her death worthy of investigation by the coroner. He subsequently released her body to a funeral home and handled the death certificate himself. The manner of death was reported to be natural causes. If it were anything other than that, it would fall under the jurisdiction of the coroner. A funeral was held for Shelia, she was duly buried, and that seemed to be the end of it.

In police jargon, a "rat" is a person who will "give up" another person in order to cut a deal. The deal usually involves getting a reduced sentence for a crime the rat has committed. Now, it's an old police trick to put a suspect into a cell with a "rat." The suspect tries to impress the rat with his exploits. The rat is impressed, encourages the suspect to spill his guts, and then the rat snitches to the cops. You'd think the suspects would catch on, but evidently many do not. How stupid do you have to be to trust secrets to another jailbird?

At any rate, somebody who was considered a "reliable" informant ratted out Jimmy Jones over Shelia's death. The rat had enough specifics to come to the attention of the homicide division. I don't know where the rat got his information. The rat claimed Jimmy had strangled the girl over cocaine. She was with Jimmy in a well-known sleazy "crack" motel in North Baton Rouge. A real rathole, where you would expect to rent by the hour. They had been doing crack for the better part of a day. The drug use was interspersed with the ingestion of beer. The rat even knew the room number and the brand of the beer.

According to the rat, Jimmy and Shelia argued over who was hogging the crack. Surprisingly, pound per pound of body weight, women can often consume more cocaine that their male counterparts. This is because women tend to have more of an enzyme that inactivates the cocaine. You don't have to be a physiologist to know this—just ask any crack head on the street.

Cocaine tends to do lots of things in addition to making the user feel good. As the run continues, agitation and paranoia set in. For whatever reason, Jimmy reportedly took off his bandana, wet it in the sink and sneaked up behind Shelia and twisted the bandana around her neck. Having done that, he tied the bandana back on his head and did the rest of the coke. I have never known an addict to willingly leave any cocaine on the table.

Jimmy then left her for dead, but somebody found her and

realized she was still alive and anonymously called 911. The emergency medical techs responded to the motel room. She was resuscitated and rushed to EKL, where she was admitted to the Intensive Care Unit, and died.

EKL is a charity hospital that dates back to the Huey Long legacy. It is a faded gray structure that has seen better days. Since it is a teaching hospital, the residents handle the brunt of the medical work and patient loads. I am not sure how Shelia's death slipped through the cracks and was not reported to us. I suspect she had several overworked physicians attending to her and the "on call" doctor pronounced her dead without knowing the strangulation history.

I first heard of her death from the police. Based on the informant's report, the Baton Rouge homicide detectives got an order of exhumation. Their next call was to the coroner's office—I was deputy at the time. Burying a loved one is hard enough; having that loved one dug up and autopsied causes even more pain. So several months after her death, Shelia's casket was exhumed under the guidance of the office and was brought to the State Crime Lab for opening and for autopsy.

As I said earlier, Baton Rouge did not have a parish morgue. Amazing—because no morgue means no chain of custody. I cannot imagine getting a death sentence with contaminated evidence, but evidently it had never been challenged. It's the way we had always done it! The office was a travesty in my humble opinion. I was there to witness it and to learn.

You never really know what to expect when a coffin lid is opened. We all brace ourselves for the worst. Fortunately, the funeral home had done a good job on her corpse. As one detective opined, "She looked fresh." I had expected the stench of decomposition, but we were spared.

She was removed from the coffin and placed on a stainless-steel gurney, which was to serve as the autopsy table. But the cause of death seemed apparent right away. As soon as I saw the embalmed corpse, I knew she had been strangled. The strangulation marks around her neck were blatant. It should be noted that bruises often appear more prominent after a body is embalmed. I was surprised the embalmer missed it. I was surprised the corpse's makeup artist missed it. I was surprised the physician missed it. There was ample opportunity to catch it.

Shelia was reinterred after the autopsy.

Much to Jimmy Jones's surprise and chagrin, he was charged with her murder. He thought he had gotten away with it. EKL got a letter from the coroner's office on reporting requirements. Although justice was belated, the victim did get a new death certificate that listed the cause of death as strangulation and the manner of death as homicide.

I learned a lot about informants from that death investigation, and that talk is (not always) cheap. The sight of the marks on the embalmed neck of the victim preoccupied my mind for a while. I get that 3-D image of her even now as I write this. *How many have we missed like this one?*

RAT DOG

I happened to drive by it today, the house on Tenth Street, in the northern part of Baton Rouge, just under the interstate. It had been gutted by fire. It won't be long until it gets torn down, and it won't be missed. At least not from what I could tell when it first came to my attention—just a typical run-down shotgun shack.

But it once housed the nightmare of elderly neglect. I entered that nightmare one night in 1998 in response to a call from uniformed police that they had a death of an elderly black female at that residence.

The house was unassuming enough as I drove up in the vehicle I had come to know as the "Green Hornet," a faded green Ford Crown Victoria—very used—with a "police interceptor" package, which basically means beefed-up suspension and brakes and a bigger engine.

When I got up to the porch, a uniformed officer briefed me. He told me there was an old lady who had died in her bed. One of her sons had found her and called EMS, who arrived on the scene and determined that she had been dead for several hours. So I was called. The son said she had been sick a long time. EMS said they knew her from previous calls and that she had significant medical problems, including Alzheimer's disease. Seemed simple, but the officer had some concerns about something he had observed on her left forehead. He also told me there was no light in the bedroom and just one bulb working in the living room. This was a bad omen. I went back to my unit and got a powerful flashlight and an AC "trouble light."

The house had only four rooms. All the rooms were built in a row, hence the term "shotgun shack." As my north Louisiana grandpa used to say: "You could shoot a shotgun right through the front door and the pellets would fly out the back door without hitting anything in between."

The living room was unkempt and cluttered, and that was where the sole light available to us was, an exposed bulb hanging from a single ceiling fixture. The couch and stuffed chair were both torn and tattered, having seen better days. What looked like an abandoned dresser supported the TV set, which was tuned to a sports channel, blaring away. It was very annoying. The whole place had the acrid reek of stale nicotine and urine.

I walked across the ill-fitted linoleum rug and peered into the next room, which was the bedroom of the deceased. For those folks not familiar with it, a linoleum rug is sort of a poor man's floor tile. It comes in a big heavy roll that looks like thick oversized plastic and has some sort of pattern on it. It is frequently rolled out over a wooden floor. In a short time, the ridges in the floor indent the linoleum and then cracks start to show up. The bedroom was dark indeed. I stood at the entrance of the room and shined a beam of light into it. There were only two pieces of furniture in the room—a bed and a nightstand. The mattress was devoid of any sheets or blankets.

I discerned some movement in the area of the head. "What the hell?" I focused my beam on the old lady's head. "Rats!" *Her head was covered with rats!* It was a horror movie come to life.

The rats were taking advantage of an easy meal. *They were eating her head!* I stomped the floor and they scurried away— reluctantly. The uniformed officer's response was limited to the same repeated phrase: "Holy shit! Holy shit! Holy shit!" We approached the bed and he pointed to the area on her forehead that was missing. There was still a telltale quiver to his voice as he continued. "It's bigger now," he said. "And her eye is gone."

I started jotting down notes in my handy three-by-five notebook, to be later transcribed on a computer. I've filled scores of them. I looked closer and fished a magnifying glass out of my "possibles" bag, which, apart from my car trunk, holds everything I might possibly need at a crime scene:

- flashlight
- backup flashlight
- mini alternate light source
- a scalpel to cut a hole in the right upper quadrant of the abdomen and into the liver
- a thermometer to push into the hole in order to get the core body temperature and maybe help determine the time of death
- gloves
- more gloves
- blunt scissors to cut off clothing (and never cut through a bullet hole)
- paper towels to sop up blood, in order to see a wound
- Polaroid camera (in case my forensic pathologist wants a view of the scene)
- reference cards for entomology (maggots and the like)
- binoculars (in case the killer threw a weapon up on the roof)

- GPS to record the exact site the body was found if out in the woods (you can drive a small metal pipe into the spot and come back years later to the exact spot with a GPS and a metal detector—vegetation grows quickly here)
- body bags of various sizes
- evidence seals
- toxicology collection kits
- insect collection kits
- boots
- rain gear
- insect repellent
- insect foggers (an exposed arm can suddenly turn black with mosquitoes down here)
- Glock handgun, 9mm (once I became coroner; Louisiana law grants only the coroner carry rights)
- extra pair of glasses
- disinfectant

I pointed out in my notes the gnaw marks left by the rats' teeth and noted the fact that there was no bleeding. All of these injuries occurred after death. They were postmortem injuries. Oddly, standing over a woman whose face had just been eaten by rats, I felt a vague relief—that she had not sustained *this* trauma prior to her death.

The officer's attention was suddenly diverted to the front room. One of the woman's sons had stumbled into the house and was yelling that he wanted to see his mother. He was very drunk and was ranting that he was a Vietnam veteran and nobody had the right to keep him away from his mother; the tension at the scene started to escalate. I was in the doorway of the bedroom. The officer stood in the living room between the drunken son and me. I gripped my flashlight tightly. It was the only weapon I had. The headline of the morning paper flashed before my eyes: DEPUTY CORONER & POLICE OFFICER KILL SON OF DECEASED. It looked like it was about to be party time when another son appeared.

Son number two was more rational and seemed willing to help with the situation. He calmed his brother down in a composed, assertive manner and introduced himself.

I expressed my sympathy to the family and told him I needed to examine his mother and take her to the morgue. He understood and accepted this. I also asked about electrical outlets and he assured me there was one in the bedroom that worked. It was on the side wall and under the bed—the bed under which the rats had retreated. I told him about the rats. I had no intention of putting my hand under there to find an electrical outlet.

This is when he made the revelation that we needed to let the "rat dog" out. Rat Dog was kept in the bathroom. He had been barking sporadically since we had entered the house. Needless to say we felt the dog was best secured right where it was. However, bowing to the wisdom and experience of the helpful son, we agreed to let him release the dog.

He opened the bathroom door, whistled, and hollered, "Get 'em, Rat Dog!" Apparently that was the animal's name. To my amazement, Rat Dog, a medium-size black-and-white mixed terrier, bolted out of the bathroom and raced right under the bed. The rats scattered as he herded them into the back room, which inexplicably contained stacks of lawn mowers.

With that, the helpful son deftly reached under the bed and plugged in my light. The whole room was immediately illuminated, as was the destitution of the situation. The woman was malnourished and had essentially died of end-stage dementia. I doubt she weighed eighty pounds. She had not been properly cared for and should have been in a nursing home.

I had seen many situations like this before. At times the person stays or is kept in the home because moving the elderly person to a nursing home would mean that the Social Security check would no longer come to the house. That check may be the major source of income for the household. Poverty has many victims—that's the harsh reality. And then again she may simply have refused to go to a nursing home, and the family accepted that as her wish.

We have all sorts of agencies to deal with such issues, but for whatever reason, the safeguards didn't work for this poor woman.

As I drove past her house again, I thought to myself: *I'll be glad when they finally demolish it.*

SECURITY BREACH

Crime scenes can turn volatile in an instant. A killer may even come back to the scene while it is being processed. It happens. Or a distraught relative may crash through the yellow boundary tape, bent on seeing the deceased and/or extracting revenge for the death. Sometimes, they just walk right in.

As I recall, it was late into the night. The time when most people are sound asleep and certainly not aware of the violence unfolding in their city. I got out of my vehicle at the death scene, and I was greeted by a veteran police officer who had been a homicide detective for many years. He had gone back into uniform due to some twist of politics within the department. I don't know the details, and really don't care to know. Bottom line: he's a damn good cop and he knows how to secure a scene.

The humidity and heat were about five clicks past sweltering. It was one of those nights when you could actually see the humidity, that smoggy haze that clings to everyone and everything. Everyone had that sweaty look. I don't mean "glistening," either—I mean plain old uncomfortable sweat. The kind that makes you as irritable as a rattlesnake. This kind of heat and humidity also tends to fog the brain. I made the mistake of asking Officer Ben Odom how he felt.

After giving me a gruff rundown of current events, he sputtered out some derogatory comments about the heat, humidity, and the mosquitoes.

Okay, then, glad I asked.

I was standing at the street boundary of the parking lot of a local convenience store. The yellow crime-scene tape had been strung up accordingly. Even at this ungodly hour, the inevitable crowd had gathered. I took my usual survey of the area and noted that there were about a half dozen uniformed officers spaced at intervals about the perimeter. "It couldn't get more secure than this," I said.

I motioned to Ben to ask if it was okay for me to go in. He gave me the nod and I entered the crime scene proper. There was no hurry, as neither crime-scene investigators nor detectives had arrived yet. But I lifted up the tape and walked in. I always had a certain feeling of control, if not security, inside the yellow tape. It's all psychological, of course, and that sense of security is often a false one.

The body of Calvin Brooks, a middle-aged black male, lay in the middle of the lot. Evidently he had been involved in some type of altercation. His weapon was reportedly an iron crowbar. His opponent, armed with a gun, had shot Calvin and fled the scene. This would become a police legend, and I could hear it already—the one about the guy who brought a crowbar to a gunfight.

Calvin was lying face up and flat on his back. He had a light tan jumper on over a black T-shirt. A coat in this heat could mean only one thing: he was up to something. Maybe he was concealing a weapon or some type of contraband.

A seasoned officer would have questioned this right away. *A coat? In this weather?*

He had sustained a single gunshot wound to the head. A pool of blood had oozed from the wound and added to the stains on the pavement. I crouched down beside the body to do an evaluation of the injury. I was examining it to determine the distance that Calvin had been from the shooter. Holding my magnifying glass and SureFire flashlight, I was looking for any sign of stippling. In close-range gunshot wounds, the powder can actually burn little speck marks into the skin. I was also looking for any unburnt powder grains. Different brands of ammunition often have different types of gunpowder. The grains can be round, flaked, or cylindrical. Needless to say, this type of search requires concentration.

A question came from my left: "They shot him, huh?"

It is not unusual for someone to be peering over my shoulder. It's usually a homicide detective. I was just doing a preliminary assessment of his injuries. I would not even touch Calvin's body until all photographs were taken and the homicide detectives had seen the scene in its undisturbed state.

I responded, "Yes, someone did, but not at close range." I then pointed to the crowbar that was halfway hidden up Calvin's right coat sleeve. "I don't think he had the chance to get any licks in. So I doubt your shooter will have any injuries. I don't see any overt signs of damage to the hands, so I don't think there was a fistfight or physical struggle with the assailant. He couldn't have swung a fist with that crowbar up his sleeve. We'll bag the hands anyway when we get ready to do a more complete examination. . . . Maybe they were waiting for him—I don't know. What do we have so far?"

When I turned to my companion, I did not recognize him. I thought I had been talking to either a detective or one of the crime-scene officers—wrong. This guy was a black male dressed in typical civilian workclothes. He seemed amiable enough, but I was immediately on guard.

In other words, he scared the crap out of me. My adrenal glands prepared me for the fight-or-flight response. *I'm in a parking lot with a murder victim and this unknown. Is he the killer, a relative, a friend, an onlooker? I'm caught off guard. The police officers are inside the tape but they are miles away if this guy decides to do me and produces a gun or a knife.*

So there we were, the two of us, stooped over Calvin's dead body.

I politely asked him, "Who are you, sir?" and followed up with, "I don't think we've met."

He responded in a calm but determined voice, indicating to me that he was a relative of the deceased. He also indicated to me that "they" knew who killed him. He seemed calm, but I have seen people go from calm to uncontrollable rage in a millisecond. Such violence can be precipitated by saying the wrong thing or making the wrong gesture. Once triggered, that person tends to strike out at the closest target. That would be me. Now, I don't know this guy, and maybe he is who he says he is and maybe he's not. He certainly has exhibited poor judgment by walking past the tape.

I knew this was a dangerous situation. I didn't know if he was armed or not. *I've got the .32 Kel-Tec semiautomatic pistol in my right pocket, but I'd have to straighten out my leg to get it. That might be a giveaway as to my intentions.*

In Louisiana, a coroner is also a conservator of the peace, which means I am technically a "police officer." At least that's what some coroners say. The corollary is that I would rather be judged by twelve than carried by six. As such, most of us are armed, and we need to be. I train with firearms, and I try to stay prepared. The law also specifically allows the coroner to carry a weapon. But it's the unexpected that gets you. It is impossible to guard against everything. I just try to use common sense and keep a cool head.

No one seemed to be paying attention to what was going on over here. I did not want to alarm this guy or escalate this into a full-blown confrontation. I needed to get him and the scene secured. I've had years of experience with talking down violent psychiatric patients—not that this guy was a psych patient. He was an unknown to me. I know that for every action there is a reaction. By staying calm, I wanted to establish a nonthreatening atmosphere. I stood up, slowly. At least now, I was standing and had a better chance if things went bad all of a sudden.

I introduced myself, and asked, "Would you mind coming with me to talk to the officer in charge?"

Much to my relief, he agreed, though he had no intention of telling the police anything. "They gonna have to find out for themselves." Translation: Calvin's relatives would take care of this if they found the killer first. And they had the advantage of knowing who it was.

We eased over to Officer Ben Odom, my old buddy, and I introduced the guy and explained the situation. Ben was calm and polite, yet firm. He asked the guy if he had any weapons on him and gave him "the look." The guy said he was unarmed and that Ben could check him if he wanted to do so. He escorted the guy out of the crime scene and had one of the rooks check him for weapons. He then stationed him with one of the other uniform officers for "safekeeping."

Once everything was relatively secure, Ben sort of went ballistic. He had what is euphemistically referred to as an "attitude adjustment" with the cops who were supposed to be guarding the crime scene. He was angry that the guy had gotten past security. Calvin's brother just walked right through. Ben Odom was embarrassed by the breach of the crime scene and furious that I had been placed in danger. He's a testy person anyway, and the heat wasn't helping his mood. His explanation and apology to me was terse and to the point: "Sorry 'bout that, Doc!"

Though it came later on in my career, it was the first—and only—tutorial I needed in the importance of securing a crime scene. The rest of the process was fairly routine after that. Detectives finally arrived, then the crime-scene unit, and we worked the area and put Calvin in a sealed and tagged body bag. At about four A.M., he was off to the morgue for autopsy at daybreak and I was heading home to take my boy to school. I try to do that at least one morning out of the week.

It would be impossible for him to understand the world I just stepped out of. We'll chat about school and stop to get doughnuts at our favorite Korean doughnut shop. And I'll just be a dad to my son and try to leave the street behind while I'm with him.

TV VS. REALITY

Frequently debates erupt among detectives and forensic investigators about how much criminals learn from TV shows such as *CSI, Law and Order,* or *Crossing Jordan.* I've heard detectives curse those shows for "making our jobs harder" by teaching the criminal how to get away with a crime. I don't think little snippets on *CSI* really make much of an impact on criminals' behavior. For one thing, criminals don't understand the foundation of the science of forensics and therefore tend to make stupid mistakes.

My experience has been that criminals generally learn just enough to botch the cover-up attempt. Do these FX forensics shows give people ideas and encourage crime? I doubt that. Attempts to manipulate forensic evidence are not new, and from what I have experienced they are not very sophisticated. In any case there are definitely thorns on that bush—such attempts often support the concept of premeditation at trial.

My wife DeAnn, son Michael, and I were having dinner at one of our favorite restaurants, Las Palmas, a Mexican place with fabulous enchiladas. We were into our first basket of chips and salsa when I got the call. I hate that look of disappointment on their faces when it comes. From a practical standpoint, we've learned to go to dinner in two cars just in case I get called away.

I ended my phone call and made the announcement: "Somebody burned up in a hotel room. I gotta go."

They both seemed somewhat resigned to it. De admonished me to be careful and Michael said he would get my order to go. We exchanged "I love you's," and I was off.

All I had at the time was: "Somebody burned up in a motel room." But as the events of the evening unfolded, the plot sickened.

It was a warm October night in 2002 when I pulled into a motel on Airline Highway. The police, detectives, fire department, and arson investigators had beaten me there. I was waved through a parking lot of fire department and police vehicles. My circuitous route ultimately led me to the second floor and the detective. He didn't look very happy.

We exchanged the usual amenities, then he got right to it. His voice was a little hoarse as he related the story: "Fire department is wrapping up. We can get in shortly . . . here's the deal so far. This may be a suicide. The guy who lives in the room had been talking suicide. He's supposedly depressed 'cause he's going to jail for some charges related to sex with a minor. So that's it right now. . . . Did we get you away from anything important?"

My response was a little on the sarcastic side. "Just dinner with my family."

His was just a terse "Yeah, me too."

This was a "special" crime all right. I was staring into the black abyss that was once a motel room—now burned out. My "street guide" was Leon Jarreau, a veteran arson investigator with the fire department. He explained: "We controlled the fire, found a dead body in the bed, and called you. We just verified he was dead, then we backed out."

I must have looked hesitant to him by the way he asked, "You ready to go in, Doc?"

I wanted to say no. Especially after his caution to watch for any live electric wires, even though "we think we have all the power off, but you never can tell in these places." I hoped he was joking, but I knew he wasn't.

This was not exactly a five-star motel. The odors drifting out of the gaping hole that was once a doorway were a mixture of burnt carpet and cheap furniture. I had smelled them before.

I also knew that the combustion of such items produces carcinogenic gases as well as cyanide fumes. Of course, the firemen had a couple of big fans going to help evacuate those fumes. I was assured that the electricity to power those fans came from fire department generators. *Now, that's good to know.*

I anticipated that I would soon be assaulted by the odor of burnt human flesh. I was not. As I entered, what I smelled was some type of petroleum. Smoke limited the illumination provided by my flashlight beam. The entire room was covered in soot. And there was that sickly squishing of the burnt, saturated carpet as I walked up to the bed that held the remains of our John Doe.

The fire had been intentionally set, and an accelerant was used, probably charcoal lighter. Since the room was relatively airtight, the fire had been limited by the lack of oxygen and had essentially smothered itself out.

There was a body wrapped in a bedspread, a blanket, and a sheet. It was totally wrapped—from above the head to below the toes. As I uncovered it, one layer at a time, I could smell alcohol. Indeed, alcohol had saturated much of the bedclothes. Strange, I thought. Maybe this was intended to promote burning. The victim was a young white male, probably in his early twenties. He was not burned. He was face down. He had livor on his back, and upon palpation it did not blanch. Strange!

When someone dies, the blood settles into the most dependent area of the body. So if you die on your back, gravity causes the blood to settle down into the skin on your back. It causes a reddish discoloration of the skin called "livor," or "livor mortis." Of course it is more difficult to see in dark-skinned individuals, but it is there nonetheless. After about eight or twelve hours, that blood becomes fixed in position. Now, once it is fixed, you can put pressure on it and it will not blanch, or turn pale—it is fixed. Also, you turn the body over and the livor will stay right where it is. This man had died at least eight hours before he was put into this position. The victim had died while on his back. Livor had formed on his back and had become fixed. Now he was face down. No policeman or fireman had touched the body. Something was amiss.

As I continued to examine the body, I noted bruises and abrasions around his neck. Then, to my surprise, I discovered the "meaning" of the crime and the crime scene: all of this guy's teeth had been knocked or pulled out. He had a mouth full of fractured dental nubs.

The entire scene had been staged for us. And it had been perpetrated by someone with the most naïve grasp of how we ID bodies. The plan was obvious. Find someone who looks similar to you. Spread the rumor that you are thinking about killing yourself and burning yourself up. Act depressed and crazy for a week or two. Get the chosen victim drunk and/or drugged. Kill him. Take out all his teeth to avoid dental identification. Burn the body up. Presto! You are now officially dead and you can safely disappear.

Wrong.

The autopsy revealed that James Page, the nineteen-year-old found in room 227 of the motel on Airline Highway, had been beaten and choked to death before the room was set on fire. If the victim had been alive at the time of the fire, there would have been soot in his nose and trachea. There was none present. He died of asphyxia secondary to strangulation. The police arrested Shawn Thompson of Denham Springs, which is just over the river from Baton Rouge. He was charged with first-degree murder and aggravated arson.

Shawn Thompson, a confessed drug dealer, pleaded guilty to the murder of the Texas teenager and signed a "Contract for Life" plea agreement that put him away for life without any right to appeal. The victim's family agreed with the plea in order for them to try and get back to some normalcy without having to go through the ordeal of the trial process. With the signing of the plea, Thompson's murderous scheme was revealed in district court.

The motive for the murder was for Thompson to escape arrest for indecent behavior with a fourteen-year-old girl whom he had been accused of having a sexual relationship with. He planned, set the stage for, and then carried out a "fake" suicide.

As the story unfolded in the courtroom, it was apparent that we were right on target with our reconstruction of the crime. Thompson had told several friends and acquaintances that he was depressed and thinking of killing himself. His "suicide" plan included slitting his wrists and setting himself on fire.

His next move was to kill his "look-alike." That was James Page, whom Thompson knew from the motel. He beat James then strangled him. Once this nefarious part of the plan was accomplished, Thompson pulled James's teeth out. He thought that once the teeth were removed and the body burned, there would be no way to identify the victim and everyone would assume that it was Shawn Thompson who was dead.

James's body was then moved onto the bed, wrapped in a blanket, doused with lighter fluid, and set on fire. The room in which Thompson set the fire was next to the one he occupied.

Police discovered the victim's teeth and the pliers Thompson used under the burners of the stove in Thompson's room.

When the body was first discovered, his roommate, Ronnie Atkins of Maurepas, told police that the dead man was his friend, Shawn Thompson. But later he gave another version of events. Finally, police matched the DNA recovered from under the victim's fingernails at autopsy directly to Thompson.

Thompson's accomplice, Ronnie Atkins, pleaded guilty to accessory after the fact to first-degree murder and could get up to five years in prison.

Few people would argue that the wave of crime shows on TV hasn't influenced the criminal mind. Tonight, I got a full dose of criminal thinking, straight out of Hollywood. What's hardest to register is that the only "crime" of this dead boy was that he looked somewhat like his killer. Talk about a random victim! It's scary as hell to think that some sociopath is prowling the city, looking for someone who resembles him so he can murder and then torch him. How do you protect yourself or your family against that sort of behavior?

I'm not sure what I'm most angry about, because I'm angry about all of it. An innocent young man is dead—for nothing. The killer is an idiot who must think we are morons. How stupid can you be?

It is extremely difficult to burn a body up—really hard.

Anyway, the fire self-extinguished. The root canals still in the jaw would have been enough for ID, and there are other means we could have used. All I need is a little DNA and your hairbrush to make a comparison.

The *CSI*-style "cover-up" failed. The murderer is behind bars. What a dumb-ass.

FOUR

What Are
the Odds?

Sudden death is always hard to explain to survivors. It's particularly hard when the circumstances surrounding a death are unlikely. One gunshot victim, a barbershop customer, was killed by an assailant who had come to murder the barber. Another man was shot while visiting a friend's home. I am sure that none of these folks got up on that fatal morning and thought, "There's a good chance I am going to be killed today." None of these people were involved in high-risk behaviors. They never saw death coming. Why does death seem to single out that person when the odds are so high against it? It sounds so hollow to say, "Random tragedies just happen." But they do.

LOVERS UNTIL DEATH

We've all done stupid things as kids or young adults. At least I certainly have. How many of us have looked back at some dangerous behavior and wondered how or why we didn't get killed doing it? Worse, as we get older, we worry that our kids may do the same stupid things, so we fear for them and we hope that they, too, will survive.

Those thoughts came flooding back to me on July 19, 1999, as I stood looking into the garage at Corley Drive. Before me lay Latonya Hall, a twenty-one-year-old female, who was face down on the concrete. Her arms were at her sides, indicating that she did not even try to break her fall. That meant that she was probably unconscious or dead at the time her body impacted with the concrete. The reddish coloration of her skin suggested carbon monoxide poisoning. Blood tests the next day confirmed my initial clinical impression.

Her friend, Jessie Hebert, also age twenty-one, was still sitting in the car; he was in full rigor mortis. Both had died of carbon monoxide poisoning.

Evidently they decided to take advantage of a friend's enclosed garage in order to have some private time for themselves. The engine had been left running to keep the air conditioner and the music going in the car. It was a fatal mistake. Carbon monoxide is a poison that sneaks up on you, displacing oxygen in the bloodstream. In some cases it causes drowsiness, headaches, or confusion, but by the time you realize you're in trouble, it's usually too late because you're incapacitated.

I suspect that Latonya may have realized what was happening and tried to get to fresh air but was probably confused and then overcome by the poison. When I arrived, the ignition key was in the "on" position, but the car engine had stopped.

My friend Stan was with me that evening and he helped me extricate Jessie from the automobile. It was difficult due to the rigor mortis. Jessie was essentially molded into place.

It only takes a body a few hours, especially in the Louisiana summer, to decompose to the point that gases build up internally and are subsequently expelled. Unfortunately for Stan, he was face to face with the body when a gush of air escaped through the dead man's mouth, and Stan caught it full force. The look on his face was a mixture of disbelief and horror. It impressed me so much that I still have a vivid visual of his expression. But he held his ground, as well as his end of the body, and he and I managed to get Jessie Hebert into a body bag. Such is the lot of a volunteer.

Accidental carbon monoxide poisoning is not that uncommon, especially after the power goes out in the wake of a hurricane and people use gasoline-powered generators as a source of electricity. One of the problems is that in some areas they keep the generators inside the house to reduce the risk of the units being stolen. Not a good idea. This time it was just kids being kids.

DEATH BY BATHTUB

As I was signing a death certificate in the fall of 1999, I paused and asked myself: How could this happen?

Gloria Reese, a fifty-five-year-old female, goes into the bathroom to take a bath on a Monday morning. This is her routine and she has done it innumerable times before. Family members note that she is in there longer than usual. They call to her. She does not answer. When they finally go in, at around seven A.M., they find her dead in the tub.

Cause of death: electrocution.

An electrical short in the stove was grounded to a pipe that touched a water pipe under the house at Connell Street. Water faucets could be used without a problem as long as the user wasn't also touching a conductive surface that would then complete a circuit through the user's body.

The woman's daughter actually discovered the problem when she kissed her dead mother before her body was removed from the tub. It shocked her. Emergency workers including my deputy didn't get shocked because they wore rubber gloves while removing her body from the tub.

By touching the water faucet in the tub at the same time Mrs. Reese had a foot in contact with the drain, she completed the circuit to ground through her body—death by bathtub.

Mr. Reese, who is disabled because of a stroke, said his wife was a "strong Christian" who got up early that day to bathe and then take someone to work. He heard her yell when she got shocked, but she was dead by the time he got there, he said. Ultimately, he had to move from the house.

Subsequently, I had our own house checked for similar problems, given that it was built some eighty years ago. It was discovered that we did indeed have a similar situation with our stove. Somewhere along the way, some genius had actually grounded a wire to the kitchen sink drainpipe! If someone had been touching the stove and the sink at the same time, that person would have been electrocuted. "That person" could have been any member of my family.

I wonder how many other older homes in Baton Rouge have the same potential death traps? *You never know.*

A MOMENT IN TIME

John Sullivan, Jr., was the well-liked and admired manager of the Louisiana State University Ag Center Livestock Show, a position he held since 1995. He had served at the LSU Ag Center since 1966, and was a beef-cattle production specialist credited with a number of innovations, including the use of computer programs for livestock shows across the state. He was a native of Zachary and a veteran of the U.S. Navy. People called him "Doc" in deference to his Ph.D.

When I first saw him one Monday morning in late May of 2000, he was face up, lying in a pool of blood about his head just inside the Ag Center Barn. Apart from the blood, he looked like he was asleep.

The LSU police, a professional group I have always found to

be on the ball, were already on the case. Lots of people discount campus police as rookies or rent-a-cops; that would be a mistake with this bunch.

I learned from them that earlier that morning, at around nine A.M., Sullivan nudged open a large metal door near the livestock office and stuck his head partially out the door, perhaps to investigate a strange sound. At that exact moment, while Sullivan's head was between the door and the door frame, another university employee, spraying weeds, hit the door with the golf cart he was driving. The driver was shaken up when I saw him, and utterly remorseful. My autopsy of Doc Sullivan would indicate severe head trauma.

I wondered how many times in the past he had opened that door and looked outside. What a tragic and bizarre coming together of events it was that caused his death.

Around the scene I heard some people say, "Stay ready to meet your maker," and "When your time is up, there is nothing you can do about it." I'm not going to take a position on the subject of predetermination. Things like this just make me realize how much of a gift life is, and how easily it can be taken away.

FOURTH OF JULY

Some people are killers and they don't even know it.

One Fourth of July night—it was 2001—the office got a call that brought me to the home of a distraught family whose nine-year-old boy had suddenly died before their very eyes.

Earlier that evening, at around 8:45 P.M., the boy had been with them in the driveway of their house.

Police say Seth was standing in his front yard on St. Gerard Avenue with his sisters. They had heard firecrackers going off in the area, lit by people celebrating the holiday, and wanted to be outside. Suddenly Seth drops to the ground. The family thinks he is just kidding around. He doesn't respond when they tell him to get up.

The sisters notice blood coming from his head and realize he is unconscious. He is transported by EMS to Baton Rouge General Medical Center, where he dies.

Seth died from a gunshot wound to the top of his head. During autopsy, the bullet was recovered and given to police for testing. Where did it come from?

The police had a ready answer. On the Fourth of July, some people adhere to the dangerous and illegal custom of firing guns into the air. Well, those bullets, after traveling vertically about a mile or so, come back down to earth. What are the odds that in all of Baton Rouge it would land on this child's head and kill him? I have no idea. This is the kind of death that can really cause impotent rage.

Ballistic tests showed it was a .380 caliber, likely fired from a semiautomatic pistol. Police then narrowed their search to a quarter-mile radius and began knocking on doors. To date, no one has come forward, and no suspect has been named. It was an accident, although I am sure a case could be made for negligent homicide. The killer doesn't know he or she is the killer. Where do you focus the anger and the blame?

DYING STUPID

When I saw Milton Graham hanging dead from a shower-curtain rod in the spring of 2001, I had two immediate reactions: (1) I was amazed that the rod would hold him; and (2) I was even more shocked by his *accoutrement*.

He had on all the trappings of autoerotica. The noose around his neck had an escape loop, which meant he was no novice at this. Duct tape had been used to hold a washcloth around his neck to prevent any chafing or noose marks. He had duct tape over his mouth, too, though his nostrils were clear of any obstruction. His hands were tied, but in reality the rope was looped about them and he could easily have escaped the bond. He had several small superficial cuts on his buttocks from an Exacto knife that he had close at hand. He was nude with the exception of a bright-red "specialty" garment that looked like some type of underwear, except for the gold metal clip that constricted his exposed genitals. Electrical clips had been clamped to his nipples. He obviously did all of this to obtain sexual gratification, and he certainly did not expect to die from the practice. But there he hung—dead.

He was hanging in the sparsely furnished room number 17 of a cheap, worn-out establishment on Airline and Sherwood known for cheap rooms—as you know, it was not my first call there. The Shawn Thompson case took place there. The pool was green with algae, and the room was what you might expect for thirty dollars a night: flat-green walls, mauve carpet, and the unmistakable aroma of stale, mildewy air.

A well-traveled path of beige and brown led to the back of the room, where the bathroom was located. By the sink we found rocks of crack cocaine.

Hanging can be a homicide, a suicide, or an accident. It most assuredly is *not* a natural death. This particular hanging appeared to be an autoerotic misadventure. I got that "crawling" sensation on the back of my neck the second I walked through the door. I think it's a primal response that alerts one to danger or fear. The whole place seemed to say, "Something is very wrong here." The only thing missing was one of those yellow signs from a freshly mopped floor that reads: "Caution/ Cuidado."

Herein lies a possible pitfall. If I get caught up in the bizarre aura of this death scene, I might jump to some sensationalist conclusion—this could all be staged. Even an autoerotic death is a murder until proven otherwise. The deceased is not some dumb freak who got what he deserved and died stupid. He is a victim, albeit a victim of his own hand.

The two beds in the room were still made up with faded blue bedspreads. The thought of what you could catch by sitting on one of them was scary to contemplate. Strewn along the brown-colored path in the carpet were several pieces of evidence: an empty plastic bag from Lowe's, the hardware store, along with several small clear-plastic bags that had been ripped open and emptied. The person opening the little bags must have felt a sense of urgency or impatience. Maybe the person was just excited and anxious to put the contents of the little bags to use. A receipt on the scene listed the items in the room: clothesline rope, electrical clamps, and duct tape—all purchased that same day.

There was no evidence of forced entry. The maid had discovered the deceased. She used a passkey to enter the room. She ran out of the room the second she saw the deceased—wouldn't you?—and called 911. I was subsequently called by EMS dispatch.

I'm not really sure what I thought or felt when I saw this man hanging. The peculiarity of the crime does not impede us from doing a thorough investigation, but it does establish a strange atmosphere in the room. Yet, having said that, auto-erotic deaths usually present bizarre scenes and elicit varied feelings and comments from investigators.

Once the initial vibe leaves the room, unsolicited bantering, snickering, and off-color comments fly freely. The veteran detectives were complaining about the weird asshole who had to kill himself on their shift. Then there were the rookie police officers who wanted a close look. And of course, the "clever remarks" were being fired back and forth:

"Hey, Tony, where do you guys buy underwear like that?"

"Hey, Frank, you know this guy?"

"I didn't know those clamps were really marital-aid devices."

"The vice squad needs to do a raid at Lowe's."

Real funny, assholes. It reminded me of a group of young boys using bravado to prove that they are not gay. I think it is a way of distancing from the victim. You sure don't want anyone thinking you can empathize with this guy.

The evidence indicated this was an accidental death—he just went too far this time. I'm sure the rocks of cocaine had something to do with his impaired ability to come back from the edge. I wondered how the family would take the news.

Is he married? What will his parents think? How will we break this news to them?

Looking back, it still seems sad and bizarre to me. It left me feeling sort of hollow. I guess I'm just not supposed to understand some of the things I encounter. But it's a strange, strange world we live in.

TRIPLETS

It was about four P.M. on a Friday afternoon in the summer of 2001 when I was summoned to a crime scene on Comstock Avenue, behind the Sam's Club store, which sells cereal, soda, and practically everything else in large volume. I've been to that store often, usually trailing behind my wife, on a quest for bargains. I know the area well.

I was informed that one of our citizens had discovered three human embryos that had been dumped out onto the driveway of a local business. EMS had come and gone. The report I got was that since the fetuses were dead, EMS had deferred the whole matter to the police and the coroner. This was the right thing to do. The initial thinking was that this was the work of an abortion clinic or worse. I was duly notified of the three human fetuses in the zip-lock bag. The call sounded most bizarre, even to me, after nearly twenty years in a big city.

Upon arriving at the scene, I noted that there were several uniformed officers there, as well as a crime-scene officer. They were all gathered around the contents of the aforementioned bag. Some were distraught if not outraged at the perpetrator of such a crime. Obviously this was the work of a heartless, perverted individual.

Even before I knelt down to examine the contents of the bag, the mystery was immediately solved. But I wanted to see just how far this had gone.

I stooped over the bag and asked, "So EMS came out here?" *Pause.* "And they looked at this?" *Pause.*

"Yes, they did."

"And they said to call me?" *Pause.*

"Yes, sir."

"Okay," I began solemnly. "Let's get some photographs as I go through this process."

I removed the first body from the bag and laid it out for a photograph. No one seemed to have a clue yet. I was later provided with detailed photographic documentation of my every move. I would have continued but my back was really starting to ache and it was really hot out there.

I could bear it no longer, so I made the announcement. "Well, I guess I have an advantage over you city folk, me being a north Louisiana boy and all."

Their brows furrowed.

"Whaddya mean, Doc?"

They awaited my explanation.

"I mean that these are huge testicles. Not of human origin.

These are bull testicles, evidently packed for sale. They were destined for a culinary fate."

"You mean they're *Rocky Mountain oysters*?"

It settled in for a moment and then the comments poured forth. It's best that the content of those remarks die at the scene.

I was told later that there was a specialty meat market in the area. I guess some unsatisfied customer had second thoughts about bringing the testicles home and elected to deposit them on the driveway at Comstock.

However, I was pleased that my earlier years of working with cattle had finally paid off on the streets of one of the most violent cities in the nation.

I know bull balls when I see them.

WITCH

"Witch"? Did he say . . . "witch"? I paused for a moment and then decided to ask for clarification. "We must have a bad connection here, Dispatch. I thought I heard you just say a witch was dead on Ninth Street."

It was true. All sorts of visuals with subtitles raced through my mind. *Black witch . . . white witch . . . bad witch . . . Wiccan . . . satanic. Wow! Just when I get to thinking that I have somewhat of a handle on this job, the bizarre reaches out and taps me on the shoulder. A witch.*

I turned onto Ninth Street at about nine P.M. The house I

arrived at looked like something out of the movie *The 'burbs.* All the houses on the street were circa 1940, nicely painted, with manicured lawns. Her house stood out. There were overgrown banana trees all over the narrow front yard. The sidewalk was overgrown with shrubbery, and the concrete was cracked in numerous places. The three steps leading to the porch were stained by algae. The porch itself was huge but darkened by the broad banana leaves that crowded it on all sides. The floor of the porch had been painted at one time but was now showing areas of exposed and rotted wood. *Watch your step, Lou.*

Of course, the porch light was nonfunctional. There was a closed screen door but the main front door was open. The creepiness was so exaggerated. I was half expecting one of the Munster clan to answer the door.

My initial encounter with the detective on site yielded a prophetic statement, "You ain't gonna believe this crap, Doc."

He was right. The word "creepy" kept coming to mind.

I entered the living room, the walls of which were adorned with mounted animal trophies, albeit worn by time. The "chandelier" did not respond to the light switch. The fixtures and walls were replete with cobwebs that accentuated torn wallpaper of a bygone era. *You've got to be kidding. This definitely feels like a bad B movie.* The floor was mined with dog crap. The flashlight illumination added to the effect.

The man who discovered the body currently had the dog in his custody. Of course, he was also the "prime suspect" of anything and everything at this time. Luckily, my path led away from that room and down a very dark hallway. The house smelled of incense—at least I think that's what it was. The whole hallway was lined with bookshelves full of paperbacks. There were rafters in the hallway with large eyebolts inserted into them. I would later discover that these were engineered to suspend ropes and chains from them.

To my right was a faint glow of yellow light. It was the bedroom and the location of the "witch's body," according to the officer. This room had the only functional light in the house—one bulb perched precariously atop a flimsy lamp that was sitting on a crowded nightstand. The lampshade was nowhere to be found. The nightstand was on the far side of the bed. The only path to the lamp was to walk onto the bed.

The bedroom was filthy. Clothing, pillows, blankets, record albums, and innumerable containers filled the room. It was dark and musty despite the incense.

What I saw next shattered any drama that might be associated with the house or its reputation. This is the real scene. She was on the bed, clothed in a cotton "peasant" dress that reminded me of the 1960s. She had her legs crossed and was leaning forward, head face down on several pillows.

One of the things I have learned is to not go rushing in to examine the body. First, look around. Once you touch or disturb anything, especially the body, the scene is changed forever. You do not have the "luxury" of morbid curiosity. You are a professional and must remain professional.

So, this is the alleged witch. She doesn't look like a witch, but then again, I'm not sure what a witch is supposed to look like. I assumed she was of the Wiccan sort as there was nothing to indicate satanic worship or the like. She was on an iron bed. Above the head of the bed was a two-by-eight-foot board that ran the width of the bed. It was tricked out with devices used for bondage and S&M sex.

As I looked about I realized that all of the containers—and there were at least fifty liquor bottles, jars, and jugs in this room alone—were filled with some type of liquid. It turned out to be urine—gallons of the stuff.

What the hell is going on here? What could possibly happen next?

In the dim light, we discover some Polaroids of some guy strapped to the two-by-eight apparatus, apparently getting his ass spanked.

This is beyond the edge! I look down at the pictures again. *Won't his momma be proud?*

Her body is stiff and cold. I have to stand on the bed to move her. There is that much stuff on the floor. The sheets and pillowcases are a dingy brown color. This is one of those times when I start to worry about blood-borne germs and other nasty little bugs that can infect a person. I worry about getting stuck with a needle or sharp object. I worry about lice and fleas and insect vectors that could carry disease to me. I worry about bringing pathogens home to my family.

I make a mental note: *Take shoes off before going into the house tonight. I will Lysol them down in the morning.*

She was twenty-three years old. Her face had conformed to the pillows. It was flat, and drool dripped from her mouth. *How sad. This is somebody's daughter.* Under her was an empty vodka bottle. A syringe was still in her left arm—heroin. "Poor girl. How did you get into this mess? You got into using drugs, and eventually the drugs used you up." Yes, I talk to corpses. It grounds me. It allows me to conceptualize scenarios better.

I examined her body for any other signs of trauma. There was none. This looked like an accidental overdose to me. She just got too deep into the pit of addiction. Young addicts can quickly develop a very high tolerance to the drug effects. In order to get a new high, they have to use in greater volume and with more frequency. It is a merciless spiral of compulsion and loss of self. Personal values, will power, and even moral choices become shrouded by the specter of drug dependency. You are lost. You become spiritually bankrupt. You see no way out. You enter into a state of oblivion. You die.

The scene reminded me of a line I read once: "One of the things wrong with her is that she doesn't know what's wrong with her."

I spread a white body bag and a homicide sheet next to her on the bed. That sterile white sheet was such a contrast to the rest of the bed and bedding. I did not think this was a homicide but I believe in doing the full-court press every time. *Once you touch or disturb anything, the scene is changed forever. You can't go back if you miss something the first time.*

In order to get her into the bag, I had to break the rigor, the stiffness, of her muscles. In other words, I had to straighten her out. She was light and not muscular, so her muscles gave way rather easily. Her full rigor mortis meant that she had probably been dead for at least twelve hours. I zipped up the body bag. She was now ready for transport to the morgue. Once there, I could do a more thorough examination with proper lighting. I would also order additional forensic studies, including toxicology and rape tests. *What if someone shot her up?*

After that, I began to search about with the detective for any additional clues that might explain what brought a twenty-three-year-old girl to this type of death. As I mentioned, there were no other functioning lights in the house. With the assistance of a flashlight, I wandered down the remainder of the hall. I was leery of something jumping out and biting me—a wharf rat, for instance.

I walked into a room that was clearly once a kitchen and stumbled upon a key clue to her final moments. On a kitchen table, I found a handwritten spell book. Several of the spells called for human urine. Well, that explained one thing. I guess she planned on casting lots of spells.

That was about the extent of my findings.

Cause of death: respiratory arrest due to drug overdose.

Manner of death: accidental.

Responsible parties: the victim, the ones who used her, the ones she used, the drug dealers, the failed war on drugs, and hence the government, society, genetics. There is always plenty of guilt and blame to go around, and in the final analysis she is just as dead. The answer, as usual, is all of the above to varying degrees.

It is frequently said in my field that the dead have much to teach us. I have a passion for the study of all types of religion. Wiccan philosophy and practices involve pagan rituals, but from what I gather, a big part of it is about doing good, not evil. Wicca is certainly not my cup of tea, but it is not my place to judge what spiritual path another person chooses to follow.

It is my place to apply scientific principles to ascertain facts related to the time, manner, and cause of death.

It is important not to judge others, a weakness that can distract me from my duties and the forensic process. Any prejudicial bias can taint the investigator, and hence the investigation. If that happens, I betray not only the dead, I betray myself.

Yes, it is frequently said that the dead have much to teach us. Sometimes, they teach us about ourselves.

FIVE

Too Young to Die

BON MARCHÉ MEANS GOOD-*BYE*

One day in August in my first year as coroner, a call interrupted lunch at home with my youngest son, Michael. There wasn't a cloud in the sky, and we had planned to spend the day together. Lunch—and unfortunately, Michael—would have to wait.

I arrived at the scene within five minutes of notification because the mall is so close to my home. It was high noon. "Mall City" is a section of Baton Rouge that had become synonymous with drugs and death. The area is well defined and takes its name from what was once Bon Marché Mall (in French, *bon marché* means "bargain," or "good buy"), a cavernous structure housing a movie theater, restaurants, department stores, and specialty boutiques, all facing Florida Boulevard. Unsupported, the mall was now essentially abandoned—despite several attempts to salvage it. However, it does seem to keep the police precinct stationed there busy. There is another salvage attempt going on now. This one may actually work!

Florida Boulevard is a demarcation line of sorts between north and south Baton Rouge. North Baton Rouge is considered a high-crime area. Behind Bon Marché Mall lies Mall City. The streets there are named after such artists as Renoir and Monet—names that are in sharp contrast with the decline of the area and the reputation for danger and crime that now defines it. It's a well-deserved reputation. Not many cases surprise me there. But, this one—this was a new low.

As I drove up to the crime scene, I could only think to myself: *How on God's earth does this happen?*

I stood over the body of a young black male. He was seventeen, and looked younger. His new purple off-road bicycle was lying on the sidewalk and he was on the curb. His eyes were still open, as if he were looking up at the pale blue sky. At first it just looked like some kid had wrecked his bike. I almost expected him to get up, look embarrassed, and ride off. But this kid wasn't getting up. He was dead. He had been executed by two other drug dealers, or "gang bangers," of similar age. I'll call the decedent Tyrone. He had on a white T-shirt, short pants, and Nike tennis shoes. His Raiders cap was still on his head.

I began to go through the paces of the investigation but all the while there was the nagging sensation. *This is not supposed to happen. This is the stuff you read about happening somewhere else, not in Baton Rouge. Hell, I can see Florida Boulevard from here. My family and I drive Florida Boulevard all the time.*

It's called cognitive emotive dissonance. You know it cognitively. It's a dangerous place. But emotionally you can't make the connection that it's really happening.

I thought of the people driving by right now, unaware that one block off the boulevard, two kids killed another kid in broad daylight. I reflected on how insulated we become in our own little worlds, and how false that security is.

Tyrone was a skinny kid—a skinny kid with two bullet holes in his chest. There was hardly any bleeding, which told me he had died quickly—probably a hole in his heart. I noted a tattoo on his right shoulder and asked the detective if it had any particular significance.

"Yeah. Seen it before. Gang tattoo." His matter-of-fact response was chilling. This would likely mean gang retaliation and more kids being murdered.

There was an empty gun magazine at Tyrone's feet. I could tell without touching that it belonged to a Browning 9mm semiautomatic pistol. The pistol was nowhere to be found. Tyrone had several rounds of 9mm ammunition in his front pocket. No two of the cartridges were of the same brand or bullet design. I suppose he had acquired them individually, whenever the opportunity presented itself.

The detective was also going through his paces. He had been in homicide for many years and he was familiar with this neighborhood, or "hood." He knew the drill cold. This would not be a mystery for long. Someone would talk. Hell, he figured to have this one wrapped up by sunset.

I asked the detective for an opinion about the pistol magazine. He said that one of the shooters had probably taken Tyrone's gun. Or someone else had come along after the murder and picked up the gun. "Of course, it ain't that hard to buy a stolen gun up in this area."

People were beginning to gather. The news media arrived. Tyrone was the star attraction today. I cringed at the thought of his mother being informed that her teenage boy was dead. Hopefully, we would tell her before she saw it on TV. But both the detective and I knew that the grapevine—the underground telegraph, or "the drums"—would have already gotten the information to her. As we stood there rehearsing how to break the news to her, she probably not only knew her son was dead but also who had murdered him. We also knew that it was best to move Tyrone's body to the morgue as soon as possible, because things tend to get really crazy out here when a kid has been murdered.

Even though I was green to the ways of city life, I didn't see a drug dealer shot by rival gang members, nor do I today. I see a skinny kid with two bullet holes in his chest. I see a scrawny little teenage boy dressed like any other—no gang colors, just a white T, shorts, and a ball cap. I see him often.

A QUARTER-INCH FROM FREEDOM

"We got a house fire, Doc—over on Monet—Mall City—they're still fighting the fire. . . ."

The knot starts to grow as soon as I hear the words "house

fire." I've been to too many of these, I guess—but one is too many. The knot turns to nausea as the next anticipated words reach out through the phone and slap me—"There're two children trapped inside . . ."

Shit! . . . I'm angry, and the visuals of previously burned children rush up from the inner recesses of my brain, and the haunting intensifies . . . the dead children call out, and the call echoes in my brain . . .

"We're still here, you know, and we always will be."

I pause for a moment. *Why am I doing this again?* I feel guilty about questioning myself.

It's about nine P.M. in the fall of 2002, almost three full weeks before trick-or-treaters raid the neighborhood. As I turn onto Cézanne Avenue, I see the flashing lights about six or seven blocks down. The outskirts of the crowd begin at about three blocks out. The main crowd has gathered and is five deep at the crime-scene tape barrier. It's muggy and the acrid smoke assaults me. The media is here and we exchange a wave to the tune of "Lemme know what you find, Doc."

There they are, the "usual suspects," as they say. One of my investigators has arrived and is standing by the coroner van. Hoses are stretched across the parking lot. Smoke is still seeping out of a hallway. One has but to trace the smoke to the crime scene. A cascade of water is pouring out of the building and onto the parking lot. Of course, all electricity has been cut to the apartment complex.

Should have put boots on. Nice job, Lou. Oh well. Too late now.

The detectives are in a huddle with the firemen and the arson investigators. We exchange greetings and I get the rundown. At any fire, FD sets up an incident commander. The police are there to determine if it's negligent homicide or, worse, "intentional." The detective begins: "Okay, here it is. Mom leaves the two kids alone—she says for just a few minutes. We strongly question that, of course. The place catches on fire. We have two dead children in the corner of the living room. We think the fire started in the bedroom because of the fire damage—well, you'll see for yourself. . . . Actually, we don't know if we are going to charge the mother with negligent homicide."

I ask, "Where is the mother?" Of course what I really mean is: *What is her affect? Is she duly upset? Is she loaded on alcohol or drugs? Is there a history of call-outs to this address? Is there any case file with child protection? Was someone else living in the house— a husband? A boyfriend? Does he have a record?*

"She's hysterical and in the back of one of the units. We are taking her downtown but she can't really tell us much right now," he responds.

I explained my position: "Here's the deal for me. I am going to 'post' both of them in the morning. I'll do X-rays to rule out fractures and the like. I'll be looking for fresh fractures which might indicate recent trauma, and I will be looking for old fractures that might indicate a pattern of abuse. I'll also get some blood chemistry to check for smoke inhalation and toxicology. Right now, I'll hold them in the cooler until you complete your investigation and we complete ours. We can get together in the morning to decide where we go from there."

We agree to the plan, break the huddle, and go to work.

The investigation will include all of the above and will ultimately be reviewed by the state child-death review panel. Maybe we can avoid similar tragedies.

The apartment opens into a breezeway. The door was locked when the firemen arrived. There was a large glass picture window that is now a gaping hole—probably shattered by the firemen, or perhaps by the intense heat within the apartment, as the kids were being literally cooked.

In the corner, by what was the picture window, are two small children. The older child, maybe four years old, is a boy. He is covering his little sister, who is about one or two years old. At least that is my guess. The flashlights give little detail—or too much. The steam is still coming up from the burned furniture and the water is several inches deep and black. It smells of burned wood and textiles and burned flesh.

The light beams hang in the air because of all the smoldering materials, many of them carcinogenic—part of the job. Just because the fire is out or under control doesn't mean the smoke has cleared. The FD has a huge fan going via their noisy generators. It helps some. We are standing in a couple of inches of water and wondering if some stray electrical current is going to make us the next victims.

It stinks. The smell of cooked human flesh only becomes more intense when we disengage the two little corpses from each other. No one says a word—the furrowed brows and grim facial expressions say it all. The corpses are handled ever so gingerly—a last act of guardianship over these bodies. It's about honor, respect, and spiritual values.

The little bodies are stiff and hot to the touch. It appears as though the brother was indeed trying to protect his sister. They are frozen in time, like the bodies at Pompeii consumed by Mount Vesuvius when it erupted.

The little girl is not as burned as the boy. His body and a pillow offered some protection to her. All of us want to think that he was trying to protect his sister. It's part of trying to lessen the horror for her, hoping that someone would champion her, since her mother obviously had not done so. Maybe it is the responders wanting to think that deserted kids somehow get by taking care of each other, or wanting to think that the little girl received some comfort, hope, a semblance of protection that made her death less horrific for her. Maybe it's just that everyone needs a hero and we all tend to look for one. The Hero with a Thousand Faces comes to mind. I don't know. It is what it is.

We will never really know. I can only imagine the fear and horror these two babies must have gone through. I envision them trying to get the door unlocked and then having to retreat to the corner of the room. Screaming for their mother and dying there in the corner, just a thin glass pane away from safety. *What is that . . . a half-inch? A quarter-inch of glass away from being burned alive?* Questions that will never be answered come to the forefront of my thoughts, but the big one is: *Why? A quarter-inch and they are out of harm's way.* It's like being tormented or taunted—such a violent, undeserving ending. In that split second, questions that span the time of our existence on this planet rush forth—questions of religion and spirituality. But . . . *deal with it later. Back to work, Cataldie!*

We pick them up, put them in the child body bags, and take

them to the van. There is something innately wrong with even having child body bags. That always bites at me.

We call it decompression. A term we took from deep-sea divers. When you are under so much pressure for so long, you need time to decompress before you go back into your real world or into your "normalcy." So, like a diver, you need to come up slowly and stop at checkpoints to adjust along the way back up. Our checkpoints might be going to the Waffle House to talk before going back home. It's a way of trying to leave the job at the job. In reality it just allows you some functionality. You never totally leave it—"it" being the job and some horror or tragedy you just dove into and came out of. Decompression helps, because if you don't decompress the pressure builds, and the explosion is released, accidentally, sideways on an innocent target. Let me give you an example. I have a really rough day. I walk in the door. De asks me: "How are you, sweetheart?'

And I blast her: "*What the hell do you mean by that. I'm shitty. Get off my ass!*" I just vented sideways. If you don't decompress, by talking about it and processing it, it all goes underground and comes out in suicide, addiction, divorce, gambling, extramarital affairs, fighting, and other risky behavior. Critical incident stress debriefing is a similar tool whereby responders gather up after the incident and just share what they saw, heard, smelled, felt, experienced. You don't have to talk. But you have to go. The only way to take the stigma out of a cop or fireman seeking help sometimes is to make it mandatory—that way you don't appear to be the weakest link. There is a social worker to facilitate the process.

It's about midnight when I get back home. Sleep is not even a remote consideration. I'm way too geared up for that. The household is asleep. My clothes smell of smoke and death. So do I.

I try to keep quiet, but my pacing about awakens De. She sees the pain in my eyes, smells the smoke, and just sits there with me in the dark living room.

It is particularly cold in the morgue this morning. Perhaps it's just the chill of dealing with dead children, but I won't dwell on the reason. My job is to be professionally objective and get "just the facts, ma'am."

We break the coroner's seal on the bag and the smell of burnt flesh permeates the whole building. I feel the sensation of nausea creeping into the back of my throat. I know how to fight it. Take a few deep breaths and accept the smell. It works for me—most of the time. The horrible odor doesn't go away. You never get used to it. You tolerate it.

I've bought various sprays that promised to neutralize the smell and not contaminate the evidence—didn't work. There was one case, however, when I had to put on a fireman's SCBA oxygen tank. A fat guy died at the Salvation Army. Dead for about a week. Hot room. I smelled him when I got out of my car—in the driveway of "The Sally" and he was in the back upstairs—take it from there. When I tried to move him, he had slippage and his skin came off accompanied by fluids. . . . But ordinarily I just take some deep breaths and dive in. If you blot out the smell, you might miss a smell that is a clue. Nothing works. You just accept it and do your job.

A random thought wanders through my brain: *Though I have been on the job nearly a decade, for the first time I notice the autopsy table is too big for a little four-year-old boy. They don't come in kiddie sizes. No, and they're not supposed to.*

He lies there, compliant and awaiting our examination— *my euphemism for his autopsy.*

We start: "The body is that of a four-year-old male who died in a house fire . . ."

Inspection of his little body reveals no evidence of child abuse, and there is soot in his little nose. There is evidence of thermal injury—burns—over about 60 percent of his body.

Michael Cramer, my trusted forensic pathologist, performs the autopsy. The announcement is made signaling the start of the internal examination. He pierces the stiff body with the scalpel, and begins the typical ritual of our profession—the first cut—*except there's nothing "typical" about cutting open a four-year-old child.*

The scalpel is pushed through the child's skin and across his little chest and then the "V"-shaped incision is made from the midpoint of that cut, then down to his pelvis—a familiar "Y" pattern made by many in this line of work. No one in the autopsy theater says a word. The silence is unnerving. *Even the hard-core veterans are having a hard time of it.*

A dissection of the neck reveals soot in the tracheal area and upper airways. The remainder of the autopsy is "unremarkable," *meaning he was just a healthy kid several hours ago. The knot is back in my gut.* A chemical analysis of his blood reveals an elevated carboxyhemoglobin level—that's the good news. He died of smoke inhalation. Smoke and combustion chemicals including carbon monoxide get into the lungs and cause an acute shortage of oxygen so that the person dies of asphyxia, rather than being burned alive.

A theory emerges from the FD: the fire started in the bedroom because the little boy was playing with a lighter in the apartment and set a broom on fire. The fire starts and they run into the other room. As the fire increases, smoke forms in a closed environment. The AC may be acting as a mechanical bellows. The children's eyes start to burn, they inhale smoke and start to cough. Their little airways—about the size of a drinking straw—become irritated; more coughing, and swelling of the airway; soot makes its way into the nose and upper airways; the air is extremely hot now, adding to irritation. The oxygen in the small apartment is being consumed; soon there is none to breathe. If they are still able to scream and cry, their little voices are hoarse and choked. Their shortness of breath increases their panic. They are bewildered, and the low oxygen adds to the confusion. They may seizure or spasm as they enter the final death throes. Mercifully, the lack of oxygen kills them before the flesh starts to bubble off of their bones.

He was dead before the flames got to him. I glance up around the room. *It's a tough autopsy. It tells on all of us.* We gently stitch the incisions up and put him into a new body bag for his ride to the funeral home.

I am zoned out as I stand there looking at the little body bag and "dealing with it."

The harsh, grating sound of a shelf being pulled from the body cooler elicits a startle response. *It ain't over yet. . . . The body is that of a two-year-old female who died in a house fire . . .*

In the end, we confirm that the little boy had evidently been "playing with fire." Mom had been away more than a few minutes.

The cause of death for both children: smoke inhalation.

Manner of death: accidental.

That Saturday, Mom attended the funerals of her two children with instructions that on Monday, she would go to the police station to turn herself in and be arrested for negligent homicide. She complied. I don't know the final outcome of the case.

CHRISTINE NOEL LOVE

"I thought it was a doll at first but then I realized it was real, and that's when we called the cops," said the BFI garbage employee who initially discovered the body of Baby Jane Doe, a few hours old in the back of his truck.

Who speaks for the dead? Who speaks for abandoned, murdered babies?

On a cool morning in December 2002, close to an Exxon refinery off Scenic Highway in north Baton Rouge, I gazed at the horror before me. Trying to look for clues and strategizing how to get the body out of the back of the truck, I felt those demons creeping up on me. Demons that cause errors that lose cases in the courtroom. These demons lurk within us with such names as judgmentalism and speculation. Most assuredly evil is visiting upon us this day.

In the midst of cascading negative emotions, it is the duty and obligation of the coroner to remain objective. I reminded myself of that very obligation as I stood, staring. *Stay focused, Lou—it's the only way to try to solve this. Don't go rushing in.*

The tiny infant (we would later weigh her out at 3.5 pounds on the organ scale in the morgue) was hanging by her head, from the blade of the garbage truck compactor. Her frail, helpless body was silhouetted by mounds of stinking garbage that had been collected before the discovery of her corpse. She evidently had been placed into a white garbage bag and then into a garbage can and subsequently picked up and dumped into the truck. The action of the compactor blade had ruptured the bag, which exposed this horror. We could not see the top of her head. She appeared to be Caucasian. The inside base of the truck was a slush of foul-smelling swill. Blood was now running down the child's suspended body and dripping off her toes and into the swill. I noted several syringes and needles in the garbage. Not a good sign, considering the task ahead.

In a world of such insanity, such bewildering intensity, one actually has the desire to rush in and embrace the baby. Sounds crazy, but once you are in such a situation you are changed forever. You cannot *un*-experience it. Even as I focused on the investigation, the reconstruction of the crime, and the collection and preservation of evidence, I knew I would be taking this child into my life, and she would be with me forever.

Of course, my role here was grounded in practical reality. The simple words from the crime-scene tech knocked my thoughts back into the moment, and back into professional objectivity: "What we gonna do here, Doc?"

Many a physician has wrestled with the mandate of objectivity and the need to keep emotions in check. Ironically, it becomes a form of subjective objectivity. Sounds oxymoronic, but there you have it. You cannot escape being human—and if you can, you don't need to be a doctor.

With some help, I was hoisted into the swill and gently tried to free the child. No luck. *Dammit!*

I turned to the crew of officers, detectives, garbagemen, their risk-management executives, and death investigators. They were just sort of staring past me and at Baby Jane. It was like being in a TV and looking back at the viewers—strange indeed. *Is this real?*

I muttered to no one in particular, "She's stuck . . . it's a she. We need to take some pressure off of this blade." I turned to the BFI folks, "Can we do that gently or gradually? And can we do it without killing me?"

Risk management assured me that they could, but I wanted to hear it from the guy who works the back of the truck all day. His assurance was my signal to continue. So the BFI guy had to move the blade. It was not without trepidation that I stood there while he manipulated the controls, which could easily crush me also. *This is surreal.*

Then . . . *Got her!* She fell into my hands. *So little, so cold . . .*

Autopsy revealed that Baby Jane had been alive when she was placed into the garbage bag. The blade crushed her skull. She had bled into the injury. *She was alive when the blade smashed into her little head.* That announcement quieted the whole morgue—no gallows humor here today. Here was silent outrage.

Some of the placenta was still in the bag. Not all of it, so maybe the mother had some left in her. That would mean she would have to go to a hospital—maybe. But she may well have left town and ended up at the Charity Hospital in New Orleans, known as the "Big Free." One could easily be treated there, and we would never hear about it. But if we do, we have the DNA to match the mother to this heinous abandonment.

Who speaks for the foundlings? We all do. We speak by what we do or don't do.

In 2000, the Louisiana legislature passed a law allowing mothers to leave newborns at hospitals and police and fire stations up to thirty days after birth without being prosecuted. *So what happened here?* Maybe the mother never heard about this safe-haven law. I'm not sure how many people in need would know about it. It's really not that well publicized. I posted it on my coroner website but I doubt the mother checked the site out before committing the murder of her child. I think that it is important that we explore the "maybes" here. Maybe we can find some answers and practice some preventive medicine. Baby land is too full already.

For Baby Jane, a group called Threads of Love speaks and remembers. This organization, with 125 chapters across the country, provides clothing, blankets, and other handmade articles for infants that are stillborn, miscarried, or even premature. They are also dedicated to abandoned infants. For Baby

Jane they coordinated a proper burial: Welsh Funeral Home prepared Baby Jane's body; a burial plot was donated by Green Oaks Cemetery in their "Baby Land"; and she was dressed in clothing hand-sewn by Threads of Love volunteers. The news media presented the story without sensationalism. They also took this occasion to promote awareness of the Louisiana law that allows you to drop an unwanted baby off at certain locations—no questions asked and no criminal charges filed. You can walk away without legal consequence.

Who listens? The funeral was well attended. Sissy Davis of Threads of Love gave me a picture of Baby Jane dressed in her outfit. She looked so peaceful—a stark contrast to the photographs taken at her crime scene and autopsy. I put it in her file. *It's about dignity.* They also gave Baby Jane Doe a name: Christine Noel Love. But my death certificate will still read "Unknown," and so will the birth certificate I must issue—both hollow documents.

As I stood by the casket, I spoke to Baby Jane, now Christine. "Well, 'Little Bit,' you got lots of love here today. I hope you find some peace now. We'll keep trying to do what we do, Boo . . . and that's a promise."

The priest talked about forgiveness. He also knows about justice. I realize that he, too, must live with his own brand of subjective objectivity. He spoke of the child and read the Scripture. It was a clear day in Baton Rouge. The wind was gentle and the birds were chirping. Baby Love's angels, I surmised. After the service I spoke with the priest. We stood there before the small casket, which was laid out under a giant live oak tree. I confided that I was having a great deal of trouble with the forgiveness part of his remarks. I could tell by his eyes that he knew. *Enough said.*

Sometimes in the early morning hours, when I'm out alone on my Harley, doing what I consider stress management and trying to get my thoughts together, I cruise into Green Oaks Cemetery. I've ridden Harley-Davidsons since the early seventies. It's escapism on an iron horse—with all the smells, sights, and feel of the world around you.

The wind feels good. The loud guttural rhythm of my "Fat Boy" engine doesn't seem to bother anyone at that hour. I park at Baby Land and visit Little Bit. There are usually several baby toys left by others on her headstone. *Maybe her biological mother has been here. Maybe to ask for forgiveness. Who knows?* I look around the cemetery and ponder that possibility. I guess I'm looking for suspects. *Does it matter right this minute? I think not.* I've stopped asking the why during my visits.

CRIME AGAINST NATURE

I've been avoiding making this journal entry, even though I still play it back in my mind often. I've had several false starts, and I've come to the conclusion that my block to writing it is that I possess no words vile enough to describe the crime. The betrayal and murder of an innocent child is such a dastardly deed that it tends to defy explanation. It is even more obscene when the perpetrator is also the child's parent.

Courtney LeBlanc, a twelve-year-old white girl, was reported missing on November 11, 2002. As the hours and days dragged on without finding her, we anticipated the worst. She was a resident of a neighboring parish at the time of her disappearance and as such, she was not in my jurisdiction; but she came to be. Her body was found in a wooded area near the East Baton Rouge Parish side of the Amite River. Her killer was her stepfather, Gerald "Jimmy" Bordelon, a forty-year-old, three-time-convicted sex offender with short gray and black hair, and icy blue eyes.

He confessed to the crime on November 26, 2002, while being interviewed by the FBI and the Livingston Parish Sheriff's Office. He was booked on first-degree murder and held without bail.

I was with my family on the Redneck Riviera (Gulf Shores, Alabama) the day her body was discovered. My chief deputy coroner handled the body retrieval and kept me informed of every detail during that process, but my family vacation was over. I returned to be greeted by her autopsy and ghastly crime-scene photos.

The hours in the morgue were silent, and the process isn't really what's important here. Courtney's story is.

She was a cute little girl with sparkling blue eyes and blond hair, and she is smiling in the photos that were frantically posted after she went missing. That smile concealed a secret hell—and it was a sharp contrast to the skull protruding through her decomposed face at the dump site.

The criminalist in me would call her story a crime reconstruction; the human in me calls it an abomination; and it seems a little late to have people listening now, when there were so many opportunities along the way to have interrupted the path that ended with her death.

The path her killer chose was dark from the beginning. By the age of sixteen, Jimmy Bordelon knew trouble well—the kind of trouble some adult convicts achieve in a lifetime—pleading guilty to aggravated rape and simple kidnapping in February 1979.

Two years after this release, he struck again. In March 1982, Bordelon abducted an eighteen-year-old woman at knifepoint. He then forced her to give him oral sex twice and released her. After her escape, she quickly led police to his home and he was nailed. He was arrested for aggravated sexual battery, aggravated assault, aggravated kidnapping, and two counts of aggravated crimes against nature. A few months later, he pled guilty to sexual battery, which is a felony, and the man was sentenced to ten years in jail.

After being moved to a few different facilities, though, he was released for good behavior in July 1988. He wasn't even placed on supervised parole when he got out of prison four years early!

Two years later, in June 1990, Bordelon was accused of kidnapping a twenty-one-year-old woman and raping her behind an abandoned building on Florida Boulevard—not far from where Courtney's body would be found. He "pled out" to reduced charges of forcible rape and crimes against nature. The kidnapping charge was dropped. He should have faced forty years for the forcible rape alone, but he was out in ten.

If our legal system had held him truly accountable, Courtney would never have been exposed to this sexual predator. None of these horrible things would have happened to her.

Jennifer Bordelon said she met Bordelon after she placed an ad on the Internet to sell a go-cart, which he bought. They married in July 2001, and two months later, the family moved to Gloster, Mississippi, where his parents lived.

Courtney's mother married a convicted rapist then moved him in with her children. A report from the Amite District Adult Probation and Parole Office revealed that Jennifer Bordelon was aware of the new husband's adult convictions from at least two sources. *Who marries a convicted rapist? Who indeed would be so naïve, so foolish, so desperate? I do not understand it. Who allows such evil to live with children? I will never understand it. Her mother must take responsibility for that behavior. She introduced a sexual predator into her family and he preyed upon her daughter. There are no surprises here. What did she expect? The best predictor of future behavior is past behavior.*

In Mississippi, soon after they moved, he allegedly sexually molested Courtney. One of Courtney's half sisters witnessed it on December 26, 2001. A Mississippi grand jury did not act on two separate occasions. The jury "stalled" twice because the half sister could not be located to testify, even though videotape of her testimony was made available. The sister had been moved out of Mississippi. The case was still pending at the time of Courtney's murder.

Jimmy Bordelon was not living with the family at the time of Courtney's abduction. He was evidently just waiting for his chance to get to her. Then it happened. Courtney was home alone and he struck. He entered her home and abducted her at knifepoint—just as he had done in 1982 with another one of his victims. I can only imagine the terror she felt. She knew what a monster he was. She knew that he was going to force some sex act upon her, and she was helpless.

How does a twelve-year-old defend herself against a grown man—a man who is such a coward that he forces her to comply at knifepoint? How does she defend herself? The answer is simple. She does not—because she cannot. That was her mother's job and her father's job and society's job—our job.

He drove her to Mississippi and sexually assaulted her. Then he decided that he would have to kill her to avoid going back to prison. That was his stated reason to detectives for killing the child.

I see a helpless, terrified child driven to the woods near the river and strangled. I imagine her tears and hear her cries for mercy. He told detectives he pushed her down, straddled her, and choked her until she lost consciousness. I feel the blackness of death consuming her as her life is strangled away. I feel shame that her little body is dumped by the river and left alone for the animals and insects—thrown away with less care than we give our trash.

At the scene, her twisted little body was face down. The harsh Louisiana environment had already started to reclaim her remains. Her flesh had rotted off of her face and maggots were feeding upon her. The rancid smell of decay permeated the humid air. I'm sure she would have been so very embarrassed. There would be no chance of an open-coffin funeral. He had denied her even that!

The search for Courtney, which would draw more than a hundred volunteers, ended with Bordelon's confession. She was accused of being a runaway. She was not. She was described as acting very mature for her age, which I believe was her cry for help. A warning bell always goes off when I hear how mature a preadolescent is. And my first reflexive question is: "Why?" Why does she have to be so mature? In this case it was because she was robbed of her childhood. It had been sacrificed to a sexual deviant. She was the victim of a father who had left her soon after birth, a mother who at some level betrayed her, a stepfather who preyed upon her, a dysfunctional judicial system that failed her, and a society that allows such atrocities to continue.

Where is the public outrage? As a society, we failed to protect this child. As a society, we would continue not only to turn a deaf ear to her cries for help, but some of us would also judge and discount her. *Courtney and all the Courtneys before her and all the Courtneys who are trapped right now are crying out to us.*

Since the original abduction occurred in Livingston Parish, my files were handed over to that jurisdiction. I hear the DA there will push for the death penalty. *What else is there?*

Bordelon escaped with a convicted killer from the Livingston Parish Prison on October 18, 2003, but was found a day later. Being dissatisfied with his public defender, Bordelon requested and got a new attorney via the Indigent Defense Board. He will be tried for first-degree murder and second-degree kidnapping of his stepdaughter in January 2006.

Jennifer Bordelon, his thirty-three-year-old wife from Denham Springs, was indicted in Mississippi for child abuse and sexual battery, on charges that she failed to protect Courtney from her thrice-convicted sex-offender stepfather. She was found guilty of felony child abuse and is out on probation for five years. The judge's admonition was damning: "Every time you look into the mirror I want you to see the anguish she [Courtney] experienced, the horror and the terror." A specific condition of her probation is that she write a 200-word letter telling Courtney how she failed her. A new letter is to be written each year on June 5, Courtney's birthday.

I hate to be so graphic in my depiction of crimes. I have done so here because I do not want to soften the horror of this death. I believe that so-called silent screams can be heard if you listen.

I still don't have a handle on this one.

SIX

Head Cases

A BRIEF HISTORY

This is Louisiana, and we do lots of things differently from the rest of the United States. Coroners here are responsible not only for death investigations but also for significant duties under the state's mental health laws. The coroner's mental health responsibilities include the issuing of orders to have people picked up for psychiatric evaluation and the actual commitment of people to psychiatric facilities. It is a significant component of the job. Some people are shocked to learn that this coroner's office deals with about 3,000 psychiatric cases each year.

The origins of these responsibilities can be traced back to the Louisiana Constitution, which has been written in both

English and French. The Louisiana Legislature of 1847 established the "Insane Asylum of the State of Louisiana" in the town of Jackson, in the parish of East Feliciana. It is still there today and we still commit people there. It was also at that time (1847) that the sheriff's office got saddled with the duty of transporting "insane" patients from Charity Hospital in New Orleans to Jackson. To this day, the police still transport what are called "coroner cases" to the psychiatric hospitals.

The judge of the district was in charge of handling "said lunatic or insane person." Insanity had a much broader definition in those days and included mental illness, alcoholism, mental retardation, feeblemindedness, and even some physical abnormalities. Just because you were declared insane doesn't mean you were really crazy.

Around 1910, the Legislature enlisted the coroner into the mental health arena by mandating that the judge had to combine forces with the coroner and the physician of the "suspected person" before he or she could be committed. A "coroner's certificate" was created, which would go through several revisions, but the judge still had the final say-so as to the disposition of the "suspect."

In 1946, the Legislature codified pretty much what we have today. A family member can petition the coroner to pick a person up and examine him for mental illness, and if the person is dangerous to self or others or gravely disabled and unwilling or unable to seek treatment, the coroner can commit him to a psychiatric facility. The person must be examined by a facility doctor and if that doctor agrees with the coroner, the patient can be kept there for fourteen days.

The true power of the coroner's emergency-commitment powers was demonstrated in 1959, when one of my predecessors, Dr. Chester Williams, committed Earl K. Long, three-time governor of Louisiana and brother of Huey P. Long, to Southeast Louisiana Hospital in Mandeville—while Earl was still governor! So, Doctor Williams may have considered the governor to be dangerous to himself or others, or gravely disabled. The governor's subsequent behavior assures us that he was most definitely not willing to seek inpatient psychiatric treatment. He did not go quietly and was literally dragged into the facility. But Earl was not deterred, and he not only ran the state from the psychiatric hospital, he kept his political machine intact and well oiled. Uncle Earl was in Southeast, which happens to be a state facility, for a little over a week when he had its administrator and head doctor fired, an altogether legal act since they were state employees and he was still the "Guv." Their replacements declared Earl sane during a short court session and he was immediately released. Louisiana is a very political place, if you didn't already know.

When I took on the job, I instinctively knew there'd be days when its two spheres of responsibility—mental health and death investigation—would overlap.

"CRAZY" GEORGE

George was a Baton Rouge "personality" and was known to many as "Crazy George." I never knew him personally but our paths had crossed, quite literally, on several occasions. George could be seen "directing" traffic on Government Street or dancing in the street and stopping traffic. He would wave, twirl about, and then bow.

People seemed amused by him and some would wave back. I think he was almost viewed as a street performer. Virtually everyone in town knew him either as "that guy" or "Crazy George."

His real name was George Patrick James; he was fifty-four, from Baton Rouge, and had served in the air force during Vietnam. George was not a street performer—he was a victim of chronic mental illness, and he deserved our empathy, not our amusement.

He functioned adequately when he took his psychiatric medications. But for various reasons he went off them and his various pathologies overtook him. Records show that he had been diagnosed as paranoid schizophrenic, antisocial, bipolar, and an abuser of alcohol. He had many episodes of violent behavior. His family had him picked up on several occasions and brought to mental institutions.

I'm sure the people in their cars who laughed at him or indulged him had no idea that he had been arrested twenty-two times for battery, assault, and even concealed weapons. People who suffer from chronic mental illness often come to the attention of the police, though. George's last interaction with the police was a devastating one.

On February 28, 2001, at around 5:45 P.M., a Baton Rouge police officer was flagged down at the intersection of Perkins and Lehman streets by a female driver. She said that George had thrown a bottle at her car. George ran into a nearby house and came out with two knives. He retreated into the house when backup officers arrived. After attempts at negotiation, an officer broke through the door of the bedroom where George was holed up. Armed with a knife, he lunged at two police officers, who shot him. He died at a local hospital. We recovered three bullets from his body.

The ripple effect of George's death resulted in turmoil for his family, the police officers who shot him, their families, and the whole community. A grand jury cleared the officers of any wrongdoing. The NAACP called on the FBI to conduct a separate investigation. State senator Cleo Fields convened a town meeting on police brutality and also represented George's family in a civil law suit against the two officers. Baton Rouge Parish settled the police brutality suit for $257,000.

I think we as a community need to assess ourselves here. There is nothing humorous about mental illness. Referring to a person as "Crazy George" is derogatory and sets up the expectation that he will act crazy. Mental health advocates bemoaned the lack of adequate psychiatric services in Louisiana to deal with an overflow of potential patients. I agree. We do not have adequate mental health funding. Many of our facilities have long waiting lists, with the result that once patients are stabilized, they are sent back out. Then they relapse and need to get back in. It's referred to as "swinging-door hospitalization," and the repeat patients are known as "frequent flyers."

I am anxiously waiting for one of our local legislators to call a town meeting on *that* issue. A host of social ills converged to cause this disaster; I'm just sorry we didn't catch them sooner, for George's sake.

I saw a man standing on Government Street today. He was dressed in army fatigues and had a red bandana on. He looked to be in his mid-fifties. He had made up some posters that were nonsensical—at least to anyone but him. He saluted me and waved at the other people who had stopped at the traffic light there. I've heard he has threatened to kill some people who live in a house nearby.

The stage is set again. Will he be labeled "Crazy John"? Will we be amused or saddened? Will we feel shame? Will we look away? Or will we find a way to help?

COMMITTED

My responsibilities look reasonable—on paper. If someone is acutely dangerous to himself or to others, or is gravely disabled and refuses to seek help, or is too impaired to seek help—I step in. Under an Order of Protective Custody, I literally order the person to be evaluated by a psychiatrist. It is my decision, and they don't get a vote. I have them picked up by the police and taken to a psychiatric facility. I have two full-time deputies assigned to me to help carry out the task: Larry Washington and Alexander Twine. They are seasoned vets and help me with many of the 2,000 cases we see every year.

After a prescribed stay in the hospital, the person is (one hopes) stabilized and then discharged to his or her home, or to some other safe place to live. Of course, the devil is in the details.

Committing someone to a mental institution is a major deal for everyone involved, and *lots* of folks are involved: the patient, the family, the office, the institution, the mental health advocacy attorneys, and the judicial system. When a request is made for an OPC, I must evaluate the motives and credibility of the request and act accordingly. If for any reason the petitioner lies under oath, there are penalties, and I'll push those penalties to the limit when that's warranted.

ANGELA: NO ANGEL

This is a composite of many, many different interviews I have done with both psychiatric and chemically dependent persons. They all tend to have the same common denominators. The patient is dangerous to himself or others. They may be suicidal or homicidal. They can also be disabled to the point of not being able to care for their needs. The patient is usually unwilling to seek treatment and the family has tried everything. Often they enable the patient by preventing him from suffering the consequences of his impaired behavior. The parents may do all the wrong things for all the right reasons. Family members are often embarrassed when they come to see me, as the coroner is often their last resort to save their loved one.

The man sitting across from my desk at 0730 hours was a man I'll call Harvey Preston. He appeared to be in his mid-fifties. He was appropriately dressed and groomed. He was relating a familiar story to me. It was about his daughter, but it could have been about any of the fifty or sixty cases the office would deal with that week. He was anxious and furrowed his brow frequently. It was as if he was pleading a case to me. In a way, I guess he was.

"The police told me to come down here and you would help me. It's about my daughter, Angela [not her real name]. She's driving me and her momma crazy, and we're too damn old for this."

I accepted his sincerity at face value. His demeanor was appropriate for the situation. He made direct eye contact and responded to my questions without hesitation.

"What is her mental health history?" I asked him.

"She lives with us. Well, off and on. She can't seem to make it on her own. We're raising her two-year-old. We're too old for that, Doc. She's had problems since she was a teenager, but it's worse now. She met some damn guy who is just no good. He got her on cocaine. We've had her to a psychiatrist and they say she's bipolar, but she won't take her medication."

Okay, there is a past psychiatric encounter. Now I've got a question about motive and secondary gain if I commit his daughter. Is this about them wanting custody of the child, I wonder. What are Mr. Preston's intentions? And where is the father of the two-year-old?

"He run off a long time ago. She gets into some bad relationships. He don't pay no child support. We just want her to get well and take care of our grandson. We're afraid she'll cart him around with her and put him in danger. I'll be honest, she's done that before. Scares the hell out of us."

That answers that. But why is he here today? Why not last week? What's changed?

"She went on a real tear last night. We had to call the police out but they wouldn't take her 'cause she calmed down. My wife really don't want her committed to some asylum, but we are at our wits' end here. We don't want her to take the baby and run off. She's threatened to do that."

I need more information about previous treatment, dangerous acts, threats to others.

"She was in a hospital a few years back and got on some medicine and acted better for a long time. Then that son of a bitch got her on that cocaine. Excuse me. Anyway, she says she ain't never going back to some nuthouse. That's what she calls it."

No excuse needed here. I let him go on without interruption. Open-ended questions tend to yield much more information in this situation. He was calmer now. This catharsis of talking it through was therapeutic for him. It is so important to have someone listen when you need it.

"She threatened to hit her momma last night—that ain't right, Doc. Then she said everyone might be better off if she wasn't around anymore. My wife is at home right now watching her sleep. We're afraid to leave her alone."

"I've got enough to issue an order of protective custody. That means I can have police officers pick her up and bring her to a hospital for a psychiatric evaluation. That evaluation must be done within twelve hours of her arrival. Now, for the real kicker—does she have health insurance coverage?"

"Nope. Not a cent. My wife and I have been talking about

this. We're willing to pay, but we ain't rich folks. . . . But she needs the help."

My gut tightens up a little. I've seen good people like this go through their life savings trying to get someone to accept help. You can't make people get well, but you can stabilize them and try to motivate them. My grandfather's words serve me well here: "You can lead a horse to water but you can't make him drink. But you can also keep him there long enough until he gits thirsty—*then* he will."

We do have a handle on this one—if she cares enough about her son. She might try for his sake, or maybe she will ultimately decide to do it for herself, and then it will stick. That's a lot of "if." Maybe I can get her a bed at the state psychiatric facility. Maybe she can stay long enough to get some good out of it. Maybe we can send her on to outpatient care. Maybe we can set her up with a halfway house for her and her kid. Maybe . . .

I explained this to Mr. Preston, with the caveat that there are no guarantees. We may send her for an evaluation only to have her released the same day. I've seen that happen. We just don't have the resources. If that does happen, I'll be sitting here again with Mr. Preston and we may well begin a swinging-door process of putting her into treatment only to have her discharged prematurely and relapse and be sent back into treatment. A coroner's hold can only keep a person in the hospital for fourteen days.

Less than twenty-four hours later, I am at the state psychiatric hospital visiting with Angela Jeanne Fontaine. Greenwell

Springs Hospital was originally a tuberculosis hospital. The inside has the look of an old institution, its tile walls a pale shade of lime. The dark finished woodwork comes from a time when craftsmanship meant something. My footsteps echoed down the shiny tiled hallway. I go there so often that I have my own keys to the locked wards. I pass through those doors and am immediately recognized.

"The coroner's here. Who all do we have for him?"

I knew it. Per usual, there were several.

One of the nurses had escorted Angela into the interview room and was kind enough to stay with us during our visit. Angela's hair needed the attention of a brush or comb. She was dressed in the standard-issue GWS hospital pajamas and robe. *Where do they get those damn things?* The robe hung loosely about her malnourished frame.

The room is not exactly conducive to establishing rapport. Lighting consists of a fluorescent ceiling fixture and whatever daylight can trickle in through the single barred window there. It has an exam table, circa 1950, with a wobbly examination light beside it.

The only other furniture in the room consists of a World War II–era battered gray desk and three chairs of various color and design, all of them chosen for their functionality. That means that if patients have urinary or bowel incontinence while sitting on them, the chairs can be easily cleaned. I'm not sure why my chair, too, is like that, and then I start to wonder about who has been sitting in it and about what previous fate it has endured.

Angela was staring at the floor when I walked in, but she

managed the energy to look at me when I introduced myself. Her eyes were framed in sunken sockets.

My obligation was to determine whether she truly fit the criteria to keep her in treatment against her will. Her responses were monotone and cryptic.

"I know who you are," she began. "My parents had me picked up. They are always trying to control me. I would have come on my own if they had asked me . . . just not right now."

I needed to explore this more. Is she unwilling or unable to seek treatment?

"Yeah. I'da come, but I don't need to be an inpatient. I got a kid, you know."

She has poor insight into this situation, and she is already trying to play the kid card. It's a rather primitive defense at this point and I confront her on it.

She tears up. "Hey, I love him! Okay? I'd do anything for him . . . except this . . . I don't need it that bad. I need to be with him."

Is she an acute danger to self or others? That is the question. "Have you had thoughts of suicide?"

"Yeah, I thought about it, but I would never really do it. I know I need to stay around for my little boy. Hell, his daddy is nowhere to be found. . . . Sometimes I do think everyone would be better off without me around. . . . I overdosed a couple of times, but I don't think I was really serious."

She's probably had suicidal thoughts. She even tried to act out on it. She must be in a great deal of emotional pain. She has the look of someone who carries a heavy burden. Her shoulders are hunched over even as she sits before me. Her face carries a seemingly perpetual frown. She is probably in direct violation of her own spiritual values. On some level she knows this. I'm glad to see the guilt. It means she has a chance. It can help her or it can kill her.

"I can't help myself at times. I know it's wrong, but I do it anyway. And I get so damned depressed, and the thoughts come . . . you know, about killing myself. Take a little too much crack or whatever. Walk in front of a truck."

The feeling of despair emanates from her. She begins to sob. True to form for a government office, there are no tissues in the room. I offer her a paper towel to absorb the tears.

"I'll stay . . . and I'll try . . ." She buries her face in her hands and begins to rock back and forth—so much pain.

Her defenses crumble but she is likely to regroup. There will be attempts at manipulation, threats, hostility, and all the things she uses to keep people away. None of it will work with this staff. They care and they know how to deal with the bullshit. But right now she is real. She is vulnerable, but this is a safe haven. She appears to be one of those women who has been used and abused. I seem to be thinking a lot about my grandfather today. A professional cowboy from Oklahoma, he would have said she had been "rode hard and put up wet." He'd consider that no way to treat a horse, let alone a human being, especially a woman.

Angela is out of control and swimming against the riptide. She is lost. But she is one of the fortunate ones. She has an anchor. Someone still cares about her. I agree with her involuntary admission. She is dangerous to herself and to others (her child), and she is at best ambivalent about treatment. But she still has a small spark of herself and her values left. I hope it will ignite. I hope we can give her a chance. I break protocol and give her a hug. She needs it. All three of us in that little dingy room are emotionally drained.

As Angela leaves the room, the nurse hands me another chart. It is with a tinge of emotional trepidation that I open it. A cursory review tells me this is going to be a very long day.

POWER OF THE PEN

The volatility of dealing with mental health patients is exemplified by events that occurred during a previous coroner's administration. A patient attacked a police officer right there in the coroner's office. The officer was in mortal danger and was rescued only when another officer shot her attacker dead on the spot.

When you are involved with as many mental patients as I am, there are inherent dangers. Some institutionalized patients know who I am, and that I am the person responsible for their current address. Some are grateful—at least while they are taking their medication—for their improved quality of life.

Others, when their symptoms reemerge, either because they are not taking their medication or because they are using street drugs, can incorporate me into their paranoid delusions. Last year, for instance, I got a phone call from a mental health clinic. It seems that one of the patients, a known paranoid schizophrenic with a history of violent behavior, had called for help. The content of the phone conversation was that he was afraid to go to sleep because the coroner sneaks into his room every night and rapes him. His solution was to eliminate the coroner. No, there is absolutely no basis to his claims. But, yes, we did manage to get him the help he needed.

YOU NEVER KNOW

One day in the early 1990s as I was leaving Baton Rouge General Hospital, a patient I recognized appeared out of nowhere. He suffers from schizophrenia and cocaine dependency, a dangerous combination. It was late and I was walking toward my car in the doctors' parking lot. Problem was, I was focused on the car and did not have my guard up 100 percent. What I did have was a .32-caliber semiautomatic pistol in my right-hand pocket. I heard a grumbling voice to my left.

"I'm gonna kill you, you motherfucker!"

When I turned I recognized him immediately. He had that wild cocaine look in his eyes. His fists were clenched. He's a big guy, and it had taken several mental health technicians to restrain him the last time he was committed to the acute psychiatric unit. He also was HIV positive. In other words, he scared the hell out of me. The last thing I wanted to do was get into a fist fight with him, but I didn't want to gun him down in the parking lot, either.

My response was automatic. "Hi," I asked him. "You doing okay, man?"

He looked surprised, unclenched his fists, and said, "Yeah, Doc, I'm okay." Then he turned and walked away—much to my relief. You just never know.

Cocaine dependency is a real bitch. The drug causes delusions, grandiosity, and paranoia, among other things. Officer Washington had escorted Arnold from his parent's pool house to my office under special invitation of an Order of Protective Custody. He was an unkempt thirty-three-year-old white cocaine addict who was determined to manipulate his way out of this little inconvenience. Since he was broke, he had resorted to stealing money from his parents and taking anything of value that he could find to hock at a pawnshop. Chief in his mind was the fact that he owed money on the street, and those folks tend to get their money or else.

At any rate, he was telling me that nothing was wrong with him and that his family was plotting against him and that was why he was in the hospital. He was most convincing, since he believed what he was saying. I could tell he felt confident that he was presenting himself and his case quite well. I committed him. But I did give him a word of advice. "The next time you try to convince me how normal you are, don't put your pants on backwards." At which point he looked down to realize he had done just that. Then he threatened to sue me for kidnapping him.

He stabilized within a few days and was discharged. He later came by the office to apologize for his behavior and to let me know he had gotten into an outpatient treatment program. I never had to pick him up again.

BEDLAM

The Hospital of St. Mary of Bethlehem, known also as Bethlem Hospital, was originally established as a sort of general hospital in London in the mid-1200s. It ultimately had its name shortened to Bedlam, at least in the vernacular. A few centuries later, it became a hospital for "lunatics," and since then the word "bedlam" has been associated with insanity and confusion.

I've been to Bedlam—the one in East Baton Rouge Parish. When I go to psychiatric hospitals for commitment evaluations, things usually go smoothly. Usually. Of course the vast majority of these patients are in an acute psychiatric unit. I should stress the "acute" part.

As soon as I was ushered through the locked doors of the acute unit one summer afternoon in 2001, I knew this was not going to be the usual. A rather large male orderly informed me immediately that "elopement precautions" were in effect. That means someone is trying to escape and you better be careful when you open any door to the outside world. I suddenly went on full alert. *Be careful here, Cataldie. If you are careless with door security, you may find yourself chasing a psychotic patient through the hallways.* I could feel the tension on the unit, and it was bordering on chaos.

The orderly has taken on more the role of a bouncer these days. I was careful as I navigated my way down to the locked nurses' station. It pays to be vigilant—one M.D. friend of mine ended up with a retinal detachment after a husky male patient cornered and pummeled him. I've been hit before. Sometimes you just do not see it coming.

I immediately sought sanctuary in the locked nurses' station, where I could review the chart on the person I was there to evaluate. The nurses' station is a safe haven, and it is brightly lit. All eyes—including mine—were on an unkempt female who was screaming through a hole in the glass barrier that encircles this station. Her red wig tilted to one side, the patient was yelling a string of obscenities at the nurse. It was as if we were under siege in a medieval fortress.

The patient, who seemed to be trying to squeeze her head through the small opening in the glass, had some white powder around her nostrils. Her anger was escalating. "You motherfuckers! I'll get you! You can't do this to me!"

The head nurse, the only nurse on the floor with thirteen psychotic patients, mostly street people, seemed unruffled. She turned to me, acknowledged my existence, then went back to the chart before her. Without looking back up, she said in the matter-of-fact tone of an overworked psych nurse: "She's a schizophrenic . . . refuses to take medication . . . we know her well . . . she's a frequent flyer. One of her thuggy sons supposedly brought her some cocaine because she promised to tell him where her government check is . . . he *supposedly* held up his end of the bargain, she didn't . . . I think he tried to fool her with talc powder . . . she may be crazy but she ain't stupid . . . you know the drill . . . it's the usual crap."

Nurse Overworked turned her attention to the screaming patient. "Listen, please go back to your room or it's seclusion for you. I don't have time for this now." Her tone was firm yet respectful. An iron hand in a velvet glove—very nice work.

This patient, Tomica, grunted another "Fuck you," then mumbled her way down the hall. She was heading toward the dining area, oblivious to the fact that her wig was leaning perilously to one side, yet remarkably still on her head. She was definitely no stranger to the seclusion room, and the threat of being put there seemed to calm her rage somewhat. Boundaries do help, even primitive ones.

I searched through the chart rack for a flagged chart that would indicate that a person was here against their will. My mandate is to evaluate patients and determine if an involuntary admission was appropriate. If I think the person needs to be here under mental health law criteria, they stay. If not, they are released immediately.

I muttered to myself as I pulled the chart from the rack and stopped at the name. *Tomica Johnston. Of course. I should have known.*

I glance at Nurse O. Finally, she smiles. This was not a nice smile. It was a misery-loves-company smile. I tried to recover. "I don't suppose you have two Tomica Johnstons here, do you?" Nurse O.'s smile widened.

I resigned myself to the fact with acknowledgment. "No? I didn't think so. Is there anyone who can help me talk to her?"

Nurse O. still had that wry smile on her face. She pointed to a young female tech who was evidently serving as hall monitor for the patients. I did not know this young lady, who appeared to be about twenty-five years old. She had probably been through the thirty-hour training course, then thrown into the pit.

As I was sizing her up for the task ahead, she suddenly bolted down the patient corridor at warp speed. "Nurse, nurse, come quick," she shouted. "Gerald, you stop that, you too old for that."

Nurse O. shot out to the hall at Mach 1. She definitely came alive when there was an emergency. I was running with her. This particular emergency: sixty-one-year-old Gerald was getting oral sex from a new admission, Alicia. They were doing the deed right there in the dining room. Alicia was on a psychotic manic swing and Gerald was somewhat demented.

The staff broke up the lovers' encounter. Gerald asked for a cigarette break and smiled. Alicia began shouting that she was going to have his baby.

Chaos is contagious, especially in a psychiatric unit. If psychotic patients perceive you have lost control or don't know what to do, they become fearful and begin acting out. They are like children in that they rely on you to furnish boundaries, and then they test those boundaries.

Nurse O. was a veteran and she knew boundaries were necessary here. We both did. Alicia was shuffled down the hall and placed in a chair right in front of the nurses' station. Nurse O. gave Alicia a stern look and commanded: "Stay right there in that chair."

The young psych tech had taken to chastising both of them.

"You should be ashamed of yourself. How people going to trust you around children?" Alicia started to sing, and would continue to do so for several hours, but she stayed in that chair.

Nurse O. returned to the nurses' station to see if there were any doctor's orders for medication to be given. There were none. She paged the on-call psychiatrist and waited. Her face was red and she was as flustered as a wet hen.

While Nurse O. fumed and waited for a return call, I elected to visit with Tomica in the hallway. It was a safety issue, hers and mine. I introduced myself and began to assess her understanding of her emergency commitment and the current situation.

She was semi-dressed in a hospital-issue gown. She had abandoned her hospital slippers in favor of her own tennis shoes—sans shoelaces. It is standard practice in acute units to take away shoelaces, which can be used for suicide by hanging. I noted that she had neglected her hygiene; there was a somewhat repugnant odor about the patient.

When she cocked her head sideways, I feared the worst for the red wig, but it stayed there, precariously perched on her head. "I know who you is. I seen you on TV with that dead baby. What you want with me? I ain't dead." There was a short pause. "*Am* I?" Another pause. "Is you the Devil?"

Never enter a psychotic patient's delusions. When I was a medical student on the third floor of Charity Hospital of New Orleans back in 1972–73, I was assigned a patient in the acute unit. I let him get between me and the door—*stupid*. Then I began my canned medical-school interview. When I asked him about his current hallucinations, he told me the Devil was talking to him right then. Stupidly, I asked what the Devil was saying. Of course, the Devil was telling this giant to "have sex with you right now." *Interview over! Now how the hell do I get out of here?*

"I'm going to tell the nurse to make sure you get double meal portions." It worked.

With Tomica, I merely assured her I was the coroner, not the Devil, and that I was there because her family was concerned about her health. Of course that was the wrong thing to say, given the recent encounter she'd had with her son.

"They don't care 'bout me."

I countered that her momma cared, and she had no rebuttal to that. It is amazing to me how many people will calm down when I say that to them. "Can you get them to give me double portions on my plate? That nurse don't like me."

She tends to switch topics easily. I asked her why she thought that about the nurse.

"'Cause the voices tell me. They want me to *get* that bitch."

There was little reason to continue. She is a known schizophrenic who is noncompliant with medication. She is experiencing command auditory hallucinations that tell her to hurt the nurse. She is acting out on those commands. I thanked her for her time but she was already distracted. She turned to her right and remarked, as if to her voices: "Shut up, bitch! Hey, Doc, I need a place to live! They gonna help me with that?"

I assured her that everyone was here to help her. She trotted off to the dining room for a snack.

I returned to the nurses' station. Nurse O. had just convinced another psychotic patient to take his prescribed medication, albeit not without some urging. I looked at her and smiled. It was a smile of understanding between two veterans who have been in this trench before.

"Tomica gets to stay," I said.

Nurse O. laughed. "Somehow I just knew you were going to say that."

I changed the topic. "So, how are the lovebirds? Engaged yet? Are we invited?"

"Ha-ha. Not funny." Her face softened a little and revealed a trace of doubt.

"You know, Doc, I think we've been in this business too damn long." She regrouped and managed a smile. "How about a cup of coffee? It's even fresh this time."

That translates into: Hey, I need to hear a friendly voice. And I need to hear it from someone who knows what the hell goes on here. I'm frustrated, overwhelmed, short-staffed, and I'm worried about these patients who are under my care. I need someone I respect to tell me I'm not crazy and that I'm doing the right thing. This sucks and I'm feeling a little on the powerless side right now, but I'll do my best. I just need a little affirmation from someone who gives a damn . . .

I smile back. *I do, but I'm a little pressed for time. I have two more hospitals to go to.*

"Sure, cup of coffee sounds good."

SEVEN

Final Exit

Some specifics of this investigation have been modified for reasons of confidentiality. The names that are fictitious include: Chris L. Cashio, Vernon Gantz, Mrs. Gantz, and Clyde P. Arceneaux. There have been at least two suicides in Baton Rouge that showed similar aspects.

A WAY OUT

I love spring mornings in Baton Rouge. The sky is clear and the temperature is in the seventies. You can smell the azaleas blooming from almost anywhere. I was enjoying this particular morning in 2000 by having coffee and warm French dough-nuts outside at Coffee Call. I love a slow cup of hot coffee, the newspaper, and time with DeAnn. We were sitting in the cast-iron chairs on the patio when my cell phone went off. I recog-nized the number as the office and announced, "Party over!"

Being coroner of East Baton Rouge Parish is a 24/7 job. De understood; she always does.

The office always rings the cell phone first, then home. De is more attuned to the home phone and I to the cell. When it rings I pretty much expect the worst. These things seem to hap-pen in the wee hours of the morning. The end of a drinking spree, fights over who is going home with whom, drunks on the road, hookers getting desperate, abusers at the end of a cocaine run. The robberies are usually shortly after dark—time to hit the mom-and-pop store. At night, after the party—time to get in the car and run down a group of kids who dissed you at the party, thereby killing five of them and becoming a young mass murderer—one of my first cases. But . . . if it has to come, I prefer an early-morning call, which is less disruptive to my family life.

The worst time is family time, in part because it induces so much guilt. "Okay, Michael, this dead person is more impor-tant than you—bye now!" I hopped into the Green Hornet and sped to the scene. I wonder if Michael will consider it un-usual for a family to go out to dinner in one car when he's mar-ried and has his own family.

As I approached the residence, I noted that there was no press, only one detective unit, and only one marked car. *Good sign.* The house sat on a corner lot of an established neighborhood. The lawn was uniformly green and well groomed. The neighborhood had been developed in the 1950s, and many of the residents were originals and now senior citizens. It was picturesque and seemed so tranquil.

Detective Chris L. Cashio was waiting in the carport; after the usual amenities, he motioned to the door and in a rather flat tone said, "This is a weird one. I think it's a suicide. Still—weird."

The word "weird," especially when uttered by a seasoned homicide detective, always bothers me and puts me on full alert. *Approach with caution, Lou.*

As I walked through the kitchen and living area I noted how neat and orderly everything was. Everything was clean and in its place. I followed Cashio's lead as he motioned for me to go through the hall and into the bedroom.

Vernon F. Gantz, age sixty-seven, was lying on his back on his made up bed. His feet and arms were crossed as if he were taking a nap. A black garbage bag, industrial grade, was under his hips. Another black garbage bag enclosed his head. He was fully clothed and even wore shoes. His shoes were polished. He wore a blue dress shirt and casual black slacks. His body was cold to the touch and had full rigor. EMS was correct: he had been definitely dead for at least twelve hours.

A glance about the room revealed that everything seemed to be neat and orderly. The garbage bag was sealed at his neck with a "twist." I split the bag open gingerly with my lockblade pocketknife so as to maintain the twist for evidence. Vernon's face had a calm look, and his eyes were closed. There was no blood apparent and no evidence of trauma. Further examination revealed that the livor distribution—the reddish discoloration of the skin that forms in the dependent areas of the body due to the settling of blood—supported the fact that he had died in bed.

On his dressing table were his insurance policy, a burial policy, and a set of directions on how to handle his affairs. There was also a picture of Mr. Gantz and a lady I presumed to be Mrs. Gantz. Next to this was an empty zip-lock bag and a drinking glass that was empty but had a fine, whitish powdery residue in it. The bag and glass were collected. I looked up at the detective and asked, "Find anything I need to know about?"

Cashio was still rather flat as he responded, "Not yet. Haven't really looked that much. Thought I'd wait for you. If this is a suicide, it's an open-and-shut case. Makes it easy on me. Looks pretty much like that one they had in the northern part of the parish a year or so ago, don't it?"

I was somewhat surprised that Cashio knew about that. That particular suicide had been handled by the sheriff's office and this one was under the city police jurisdiction. The departments rarely share information. "Yes," I said, responding to the detective's query, then paused, contemplated a moment, and added, "Yes it does, Detective, yes it does. So, let's have a look-see at his choice of reading material." I left the bedroom.

Cashio followed and continued: "I wasn't at Seminole Street, but I heard there is some kind of book on how to kill yourself—kind of a *Suicide for Dummies* deal."

I smiled to myself. *Homicide detectives certainly have a way with words.* But this seemed like an opportunity to "enlighten" the detective, so I did. "The name of the book is *Final Exit.*" It certainly is a "how-to" book on how to commit suicide. "And I'll bet you money Gantz had knowledge of it."

I walked into Mr. Gantz's well-kept reading room and inventoried the area. Near his small reading table was a publication describing the techniques touted in *Final Exit.*

Detective Cashio seemed a little disconcerted, and, immediately thinking in the terms of his profession, he asked, "So *anybody* can buy this damn book you're talking about? *Kids* can buy this crap!" Cashio certainly wasn't detached at this point. Indeed, he had become quite animated. "Say, is there a way to get the guy who wrote this book on assisted suicide?"

"I doubt it," I said. "But, of course, you'd have to ask the DA's office if you want an official opinion." I felt a need to expand upon *Final Exit* for Cashio's benefit. There was no need to have him go charging at windmills. "This book was written for the terminally ill person who wants a way out, or as the book puts it, a 'self-deliverance.' It goes through the decision process and tells you what to write and how to set up your affairs and what to take. The idea is that you go to sleep and never wake up because you die of asphyxia."

It tells you how not to create a mess to be cleaned up by whomever. That's why Gantz had a plastic bag under his butt. He was being considerate in case he lost control of his bladder and bowels.

"I'll refer to the book when I talk to his family. Folks just don't want to believe a loved one killed themselves. You know, Detective, if a person went into Tiger Stadium at halftime, pulled out a gun, and shot himself in the head with ninety thousand witnesses present, someone would be in my office within the week with a murder conspiracy theory. Family members often have a very difficult time with acceptance." I stopped, glanced around, and asked, "By the way, where *are* the family members?"

Cashio answered, "They all live out of state. We are trying to contact them. There is a daughter in Ohio. The only person this guy had much contact with was his neighbor. One thing really bothers me here, though, and that's the fact that this guy ain't terminally ill."

I added, "Well, not as far as we know. The guy who wrote the book was national executive director of the Hemlock Society."

Cashio interjected, "Now, there's a cute name, Hemlock Society. Do they mail-order hemlock out to you?"

I smiled and continued, "No, they won't mail poison to you. But they also don't support suicide in the case of emotional or financial stressors."

"Yeah? Well, somebody shouda told that to Gantz."

I agreed, "Yeah, poor guy. I need some history here. Where's that neighbor?"

Mr. Gantz's neighbor, Clyde P. Arceneaux, had discovered Gantz. The neighbor went into the house to check on his friend of over twenty-five years because it was unusual for Vernon not to get his newspaper in the morning. Mr. Arceneaux, who appeared to be in his seventies, had collected the paper, and since the door was open, he came in looking for his friend. When he saw Vernon in bed like that, he called 911.

The neighbor reported that Mr. Gantz had been depressed for many months, "since the recent death of his wife of forty-three years." He was a good neighbor. His children lived out of state and he didn't have much contact with them. He had talked about wanting God to just take him in his sleep. He had been making frequent references to "just being tired of living."

Mr. Arceneaux was one of those honest, warm-hearted, Cajun gentlemen who always have time to help their friends and neighbors. His comments were as accurate as anyone living could make, and summed up the whole case: "Me, I t'ink Vernon just don't want to be here wit'out his wife no more. You know, he was never the same after she passed. . . . You know, he didn't laugh no more, and he just moped around. But he kept his yard up, yeah, 'cause she always was proud of dis yard. He was a damn good man, dat Vernon. Maybe he love her too much, I dunno. Dat's all between him and God now."

My autopsy revealed no evidence of physical trauma. There was also no evidence of any major physical illnesses. The stomach contents consisted of numerous small pill fragments of various colors. Toxicology was positive for two sedatives, diazepam (trade name Valium), and a barbiturate. There was a small concentration of alcohol. The sedatives had been combined with the alcohol to have a much more powerful effect and induce a heavy sleep. There was also Compazine, which would prevent him from vomiting the contents of his lethal concoction.

The residue in the glass was from the Valium and barbiturate that had been dissolved in alcohol. This was the fatal slurry that Mr. Gantz ingested—one just like that described in the book.

Euthanasia, or mercy killing, is the act of killing someone painlessly, and the term is frequently associated with people who want to avoid a slow death from a terminal illness. I think it is something all of us must face and analyze. Euthanasia is a morally and ethically heated topic. Terms such as "assisted suicide" have become familiar. Is it euthanasia not to put a feeding tube in a terminally ill patient? Is it assisted suicide to abide by the "living will" of a person who does not want "heroic measures" instituted to save his or her life?

The death of Vernon F. Gantz highlighted these issues for all of us.

While my internal conflict with euthanasia, suicide, and "assisted suicide" continues, I do not judge Vernon Gantz, and I agree with Mr. Arceneaux that it's "all between him and God now."

Cause of death: asphyxiation by drug overdose.

Manner of death: suicide.

LOSS

Powerlessness is what you feel when you come to the scene of a suicide of a young mother and wonder what you could have done to prevent it. Helplessness is what you feel when you come to accept there is nothing you can do to prevent future suicides. This is the story of our journey to do something.

Suicide tends to bring out an array of emotions among first responders—prejudice, numbness, empathy. One suicide I handled in 1998 hit everyone who worked on it particularly hard.

She was in her late twenties. I'll call her Jenny. Her two toddlers called her "Momma."

My investigation started as soon as I turned the Green Hornet into that middle-class neighborhood. The sun had just started to settle over West Baton Rouge, but the soft light allowed me to see that this was a nice brick home, one that was well maintained.

I stepped through the sliding glass doors that led into the master bedroom. I noted a plastic toy truck on the carpet. A yellow dump truck to be exact, with huge black wheels—funny how those things get imprinted in your brain.

My attention was immediately drawn to the small bathroom to my right. Jenny was sprawled out on the floor. She was dressed in a previously white sweatshirt, blue jeans, and tennis shoes. Blood was everywhere in a vast spray pattern. The entire bathroom and all of its contents were covered with a fine red mist and bits of brain matter. I watched some of it drip onto the floor as I gazed through the bathroom door. I mumbled something to the effect of, "Oh my God! What happened here?"

The crime-scene officer responded numbly, "Killed herself . . . in front of her kid, Doc. That's what."

A great sadness came over me.

Every suicide is originally considered a homicide and is investigated as such. It's not a suicide until the coroner says it is. The prime suspect in this situation would be Jenny's husband— for no other reason than the fact that he is her husband. Of course that very suspicion and subsequent line of questioning would add to his trauma and grief. I've talked and listened to many people who have been left behind by a loved one who suicided. I had not yet talked to the husband, who was being interviewed.

In order to protect myself, and my family, against infections that may be transmitted through blood, I put on shoe protectors in addition to my blue gloves. I realized there was no way to avoid stepping in Jenny's blood. It was everywhere. The meaning of words like "gruesome" and "ghastly" takes on a new reality in situations like this.

Jenny was face down. She had short brown hair, what my wife would call a "pixie cut." As I turned her over and examined her, I realized that Jenny no longer had a face. Her face and the entire top of her head had been blown off. All that was left of her head was the base of an empty, hollowed-out skull with what looked like rags of flesh clinging loosely to it. *Ghastly indeed! Look what you've gone and done to yourself—poor baby. What was so bad that you had to do this?*

The evidentiary value of the fine mist of blood had not been lost on me. I realized it meant that a high-velocity bullet had caused the injury. I found a .357 Magnum revolver under her—yeah, that qualifies. That's about a half an ounce of lead traveling at almost 900 miles per hour, accompanied by exploding gases. That will demolish skull, brain, and face like a stick of dynamite—literally. *There will be no open casket for this family.*

That fine spray pattern of blood also indicated that Jenny had been alone in the bathroom when she sustained the gunshot to her head. Such a mist would have covered anyone else in that small space, and I was informed that the husband had no such blood mist on him. Though he could have changed clothes, for all we know.

Our examination reflected that the mist was uniform

throughout the entire bathroom. There was no "shadowing" to be observed. If her husband had been in there, he would have blocked the mist and that would have left a clean pattern, or "shadow," on the wall. The pattern also indicated that she may have been looking into the mirror when she pulled the trigger. *What kind of pain and despair she must have felt at that fatal moment . . .*

There is not a whole lot to do at a suicide. Once the death is determined to be a suicide, the body is taken away and the family is left to deal with the cleanup and the fact that a loved one has just taken his or her own life. The police and I close the case. It's over for us, but just beginning for the family. There will be lots of guilt, anger, blaming, and shame in the family for many years to come. And the children will later be more at risk than others to kill themselves. It is a vicious, unforgiving cycle of destruction.

Information trickled in. When she was a child, Jenny had discovered her suicided father. It affirmed what I already knew. The cycle of suicide runs in families. It can also be contagious, reaching to friends, acquaintances, and even people who don't really know the victim.

It was also reported that one of her small children had run to the bathroom when he heard the gun go off. I thought of how confused and scared he must have been—and must still be. I doubt he could understand much of this. The sight of his mother in that condition could only terrify him. Thinking about that child just about "gutted me," as my grandfather used to say.

Sadness and anger are definitely bedfellows. Anger started to assert its presence in me as I walked through Jenny's house. There was the little pallet she had made for the kids in front of the TV. Their little cereal bowls were still on the coffee table. Their meal was interrupted by death sweeping their mother away forever. There were some of the kids' drawings taped to the refrigerator door. There were pictures of the whole family— pictures taken during better times. And realizing how much she cared for them made me sad and still more angry.

Angry that death had won.

Angry that her split-second decision was unforgiving.

Angry about what these children must face, and about the stigma of suicide.

And angry at myself as I thought, "If I know all this, why the hell am I not doing something about it?" Yes, I am responsible for the dead, but I am also responsible to the living.

It was a sleepless night for me, and for DeAnn, whom I kept up for hours discussing the matter. She is my debriefer and stress-reliever at times like this. I'm lucky to have her. How many coroners have a live-in psychiatric nurse?

Jenny's death was not in vain. The next morning I called a mutual friend of ours, Frank Campbell, who heads the Crisis Intervention Center. He was a step ahead of me. Frank had been formulating a method to go to the scene of a suicide to do on-site intervention with the family. He told me that he was just waiting for a spark to get a task force going. The coroner's office was that spark. The members of the team became volunteers of the coroner's office and we launched the program on the same night that we had our first meeting. We named it LOSS, for Local Outreach to Suicide Survivors, a team made up of a suicide survivor and a crisis therapist. We were less than one hour into that seminal meeting when we got our first call. We pulled things together and were at the site of the suicide within fifteen minutes of notification.

So, what's different now? A lot. Now when I respond to a suicide, I deploy the LOSS team. Both the survivor and the therapist assist the families and help answer some of the tough questions. They help families decide about organ and tissue donation—a meaningful living legacy for those who choose to donate. They help families tell the children. They help get the blood and gore cleaned up. They stay for hours with the family in this time of crisis. They offer follow-up group and individual therapy at the Crisis Center. They do it at no charge. And it works.

I am humbled to be a part of such a program. And I am gratified that Jenny did not die in vain.

TWO ROADS DIVERGED

There is always potential for conflict between governmental agencies. This is especially true when more than one agency has authority at a crime scene. At some murder scenes there is what we refer to as multiple-agency jurisdiction. The crime scene itself, with the exception of the corpse, is the domain of the police agency handling the investigation. The body belongs to me, the coroner. The body is, in essence, my crime scene.

A homicide crime scene is usually worked from the outermost perimeter in toward the corpse. Most homicide crime scenes are geographically self-limited. This is obvious by the fact that the yellow crime-scene tape demarcates the boundaries of the crime scene.

Over the years, the police departments and the coroner's office have developed good working relationships. We are the good guys and we want to catch the bad guys. That really is the bottom line. But, friction can develop.

The kid was nineteen years old. At the time, my middle son was about the same age and was in LSU. This kid, Johnny, was also a student at LSU. Johnny had been going through what he considered some tough times. So tough that he chose to kill himself.

I first saw Johnny sitting in his car. He was dressed in preppy clothes. The car was parked in the garage of a family friend. It was an enclosed garage and Johnny had pulled in, closed the garage door, turned on the radio, and let the car idle. He died of carbon-monoxide poisoning.

There was fine soot on everything in the garage. The garage door was now wide open, but the place still smelled of exhaust fumes. His body was in full rigor mortis as we moved him from the car and into the white body bag. Since the preponderance of evidence supported suicide as the manner of death, I mobilized the LOSS team.

That was interagency problem number one. In the early days of the LOSS team, homicide detectives were less than supportive of the concept. The last thing they wanted was for someone to be tromping around their crime scene, contaminating evidence. And they damn sure didn't want the team interacting with bereaved family members. After all, who is always the prime suspect when you suspect it could be murder, *staged* as a suicide? A family member, of course.

We managed to work all of that out over the years, but this was early in the process and there is no more suspicious creature on this earth than a homicide detective.

An officer informed me that Johnny's parents had been notified and that his father was on the way. I acknowledged that message and sealed the body bag. The crime-scene photographer took a picture of the tamper-proof seal and noted the number of the tag. This is done to insure that once the body bag has been sealed, no one can disturb it or any evidence that might be on the body. It's a chain-of-custody procedure.

With that, we were ready to transport the body to the morgue for autopsy. It was just at this time that Johnny's father arrived. He was obviously very distraught, though he maintained his composure.

Johnny's father wanted to see his son one more time before he was carted off to the morgue for autopsy. One of the detectives indicated that he did not think that would be possible. The officer was technically right. To comply with this father's wishes would mean breaking the seal. In addition to that, if the father touched Johnny's body there was the whole trace-evidence issue to be dealt with. Any time additional variables or persons are introduced into a crime scene, they can inadvertently add fibers or hair to the scene or onto the body. That is extremely important in homicide investigations.

Once a body bag is sealed at the scene it is not opened again until that seal is broken at the time of autopsy. It's one of those understood, accepted, and unspoken rules. It's part of how we catch the bad guys.

I could see no bad guys here. I ran through my mental homicide checklist. There was no trauma on Johnny's body. There was no evidence that someone had killed him. He had no known enemies. He had not been robbed, as there was money in his pocket. Examination of the garage door revealed a section of burnt paint where the exhaust had hit it for several hours. Of course, Johnny was dead long before the car engine ultimately stopped. The ignition key was in the "on" position. He had a pinkish discoloration of his skin, which is a major indicator of carbon-monoxide poisoning. In short, I was satisfied that this was a suicide.

Closure and grief resolution are important for parents—there is absolutely no doubt about that. But careful forensic procedure is essential to case solving and successful prosecutions—no doubt about *that*, either. I had to make a decision regarding our immediate course of action.

Like I said, I could see no bad guys here. All I could see was a kid who made a foolish and fatal decision, and a father who would question this night for the rest of his life. He had all this pain and nowhere to put it. The kid's *dead*. The void is there—you can't scold or cajole him out of this one—this is it. *Where do you go with it?* Here's a good kid who gets snagged up on something that would pass with time, but he sees no way out and kills himself over it. And the dad and mom have all this wisdom, but it's inert, useless—the potential to resolve issues doesn't mean crap when your son's dead. Suicides are hollow scenes.

Somewhat to the chagrin of the crime-scene officer, I broke the seal and unzipped the bag. I told the detectives that I was satisfied that this was a suicide and that no harm would come from opening the bag. Besides, it's my bag and my seal and therefore my responsibility.

It was just the right thing to do.

I must say I did so with some trepidation, knowing I'd be explaining this move later to the head of homicide. But, as I reasoned with myself, as objectively as possible, there are times when you just have to make that judgment call. And you have to be a doctor, and you have to consider the concept of "First, do no harm." It would do more harm not to let this father see his son one last time. There was nothing to be gained forensically by not breaking that seal. I would put another one on before we left and we could go through our procedure again.

Maybe I was projecting because of the similarities between this kid and my kid. They could be friends, for all I knew. But this was not the time to be second-guessing myself. As Johnny's dad knelt next to his son and rubbed his son's hair and said his goodbyes, I knew in my heart that I had chosen correctly.

Robert Frost was right: *"And that has made all the difference."*

COINCIDENCES

On the bed in a room on the third floor of the hotel attached to the Argosy Casino, one block over from the government building and my office, a sixty-year-old white male lay dead.

He'd been on a bad chase—and, of course, lost. The "chase" is a term often used to describe a compulsive gambler who thinks his bad luck is due to change, and tries to recoup all of his losses by betting more heavily. The man's nine-page suicide note was reflective of his run of bad luck: lost job, lost marriage, lost hope—and now a bullet through his brain.

Casino patrons walked down the hall never knowing the tragedy that was in that room. It's always amazed me that death can be one wall away or just down the hall and you'd never know about it. Here's a guy who has blown his brains out in a fit of desperation and these people are meandering by to go over to the casino to gamble. They participate in the same behavior that just led to this man's death—although it's not the behavior, it's the compulsion. Still, it seems weird.

The room really just plain *stank!* A pool of decomposing blood, and there was lots of it, indicated that he did not die right away. The bowel incontinence was enough to "gag a maggot." When you pull back the covers and move the body, a whole new wave of odiferous particles assaults you. It's a known fact, and I've been known to announce a warning: "Get ready, it's going to really stink."

Several of the officers became sick and had to leave the room. And of course there was the hip-shot speculation offered by the detective who noted a hole in the pillow: "Look—he used that to muffle the shot!"

How stupid can you get? He COMMITTED SUICIDE, dummy! He left a note; he wanted to be found. That is the hole made by the goddamned projectile as it exited his brain and blew out the back of his skull. I don't think he was concerned about getting caught. Brilliant.

It's like children wanting to be recognized by the teacher for their observation skills and knowledge base. You hear some really stupid speculation at crime scenes—tends to scare me.

Of course hotel security was all over the place, and then there we were: four cops, three people from my office, and a stretcher out in the hall. So I had the manager assign us the empty room across the hall, and we set up a command and a staging area there, so as not to frighten the hotel and casino guests. It also served to contain some of the bullshit that gets said rather loudly at such a scene.

Meanwhile, one of the responders and I were doing what we do a lot at a crime scene—waiting—this time for the CSI to get pictures. We began to talk about suicide, and he confided in me that years ago his father had committed suicide in a hotel, a casino hotel. Talk about a flashback. Grim. He still seemed to be searching for answers that do not exist. At least, that was my impression. This revelation is also a reminder to be careful of what you say at a crime scene.

Suicide scenes are historically fertile ground for judgmental comments: "He took the coward's way out," or "Why did this SOB have to kill himself in my district . . ." You never know who is listening, and you might be amazed at how many people have been touched or affected by suicide.

This is my second suicide in as many days. What the hell is going on?

Last night this fifty-eight-year-old white male shot himself with an 1873 Winchester carbine, caliber .32-20, a rifle that helped win the Wild West. Not a good choice—he had to shoot himself three times to get the job done. How many coroners deal with gunshot wounds from a rifle that is an antique?

He was a hoarder and had lots of obsessive-compulsive–type indicators in the house. Everything was in order. Even the cans in the pantry were arranged by size, and all labels faced forward. Newspapers were bound and catalogued. Clothes were hung by type and size in the closets.

He shot himself in the bathtub. I surmise he did not want to leave a mess. He had ejected two empties and placed them in the soap dish. He did not eject the last one, as he was either dead or incapacitated by that time due to exsanguination, or bleeding to death. A huge mirror faced the bathtub. It would not surprise me to learn that he'd stood before that mirror and made several dry runs with an empty rifle as macabre rehearsals. Of course there is no way of knowing that. Some suicide victims actually fire a test shot into the wall as a prelude to shooting themselves. I guess they want to make sure the gun works. Of course such a "practice shot" always complicates the investigation. That was not the case here. There was no note. Then again, only about 25 percent of suicides leave a note.

The house reeked of depression and mental anguish. Suicide houses often do. The lighting was dim and several bulbs appeared to be burned out, in contradiction to his need for orderliness, but I've seen this before. Shades were drawn and little light could break through. The place was orderly but not clean. One of the officers sarcastically commented that the maid had obviously not come in some time. Dust was prevalent on every surface, and the carpet needed cleaning. The mail was stacked in neat little piles but much of it had not been opened. Boxes of "stuff" occupied most of the seating space in the house. Indeed, there were only two places not occupied by stacks of paper or boxes. One chair at the kitchen table was free of clutter. It faced the table where the mail was stacked. The other open chair faced the TV set. He had lots of VHS movies, all in place and on the shelves—and in alphabetical order, of course. The TV was off and there was a thick coat of dust on the screen. All of these signs promoted an atmosphere of isolation. Maybe he was agoraphobic. He was obviously in a living hell and saw no way out—except one.

The act of suicide gives us permission to invade your privacy, ask intimate questions about you, go through your belongings, form opinions about you. Nothing is off limits or "sacred" once you kill yourself. This is a murder and we are looking for the motive. But I really didn't get a sense of knowing this man. As I talked to neighbors, acquaintances, and distant family members (there were no close ones), I realized that no one really seemed to know him. He was truly alone. I had his motive.

ONE OF THEIR OWN

It was a nice day in Baton Rouge. Flowers were blooming, wispy clouds floated by against a blue sky, and a mild breeze caressed us. This type of day is one of the delights of living in this area. I stood there before the neat little townhouses that lined the street. Several police officers came over into the shade offered by three huge crape myrtle trees. The mood was especially somber, and the conversation was almost muffled. It centered around one of their own. A retired police officer had taken his life in one of the upper apartments. It felt like we were at a wake and standing outside of a funeral home.

The case belonged to the city police but the sheriff's office had also responded. There was no jurisdictional squabbling. When a fellow officer goes down, cops take care of cops. This man was one of their own. He was a well-known and respected police officer who had fallen upon hard times.

After I had waited there a while, one of the city detectives asked me to come up. His demeanor betrayed his feelings. It would be difficult to get any distance from this death.

He briefed me in an almost flat tone. It was an attempt to be matter-of-fact, but it just didn't come off like that to me. Maybe I was projecting my own feelings onto him. I'd sort that out later.

The deceased, whom we will call David Smith, had gone into the bathroom, gotten into the bathtub, and shot himself with what appeared to be a service revolver. He had left the apartment very organized. Everything was in place and everything was clean. I'd seen this before in suicide deaths. All of his important papers were organized on the kitchen table. He had written instructions for every discipline that he knew would ultimately show up. He even had one for me, entitled "The Coroner." It was essentially instruction for me on whom to notify of his death.

Police officers who had worked with him held him in high

esteem. Many identified with him. There was an impressive array of commendations on his walls.

His cause and manner of death were apparent.

One of the things I took away from that scene was the disconcerting remark that one of the policemen made. "I often wonder if I'll end up like this someday."

I still have a crystal-clear image of that moment. David Smith in the tub with a bullet hole in the right temporal area of his skull. I remember the white porcelain of the tub contrasting with the dried blood that had oozed from his head. I remember the color of the shower curtain and the rug. And most vivid is the image of that police officer. I can still see him looking down at David, and I still hear those words. It is like the whole scene is etched into my brain, in a kind of mental freeze frame.

His words alarmed me, because that level of empathy could be interpreted as his intention to follow suit. This could serve as a rehearsal to see how one's peers view such a death.

Every time I see that particular officer at a crime scene now, I ask him how he's doing. I doubt he'd tell me the truth. Therapy for depression or mental-health issues still seems to be taboo or at least tainted in that culture. But what do I know? I'm not a cop, I'm a coroner.

EIGHT

Thou Shalt Not Kill

DOG DRUNK

I continue to be astounded by some people's motives in murdering another person. This drama occurred in the northern part of the parish. I really don't have to go up there often; they tend not to murder each other as much. Tonight, obviously, was an exception.

When I recall the events of that muggy Louisiana evening in July 1994, several vivid images stand out. The first is that of the massive number of vehicles—pickup trucks, EMS vehicles, fire trucks, police cars—that I came upon as I approached the trailer park. They were erratically parked and stretched from the actual crime scene, which was the last trailer on the left. I judged the distance to be at least a quarter of a mile.

As soon as I got out of my car, my glasses fogged over. Humidity was at least 110 percent and the heat index approached that of Dante's Inferno.

The official vehicles had the usual array of strobe lights going. If you ever make direct eye contact with a bright strobe, your night vision is shot to hell for at least ten to fifteen minutes. I tried to avoid the strobe-induced night blindness as I waded through the vehicles and the clusters of people who were grouped randomly. They all replied when I acknowledged them with a nod or "Howdy," which is the culturally correct greeting in this neck of the woods.

The next visual image I remember is the organized chaos. I was initially met by an apprehensive sheriff's deputy who told me the killer had run into the woods and barricaded himself. His speech was pressured and slightly high-pitched: "He's armed and he's dug in! We're going in after him." As if on cue, several SWAT types hustled by and entered the woods.

It was one of those slow-motion moments. If you've ever been in an automobile wreck and seen it coming, you know how things tend to slow down, and how you have this unnerving feeling of calm. It's kind of a crazy feeling. That's what this was. *Hey, I shouldn't be calm, this is a volatile situation. That guy could come charging out of the woods at any moment . . . and should a gunfight ensue, bullets might be flying all over the place.*

Some of those deputies had what my granddaddy used to call that "wild-eyed look." The adrenaline does it, it's contagious. I probably looked "wild-eyed" myself. The last thing I needed was to be caught in a crossfire between the police and the perpetrator. I thought to myself that I should be wearing a vest—of the bulletproof variety—but that was not the case. I turned my attention to Raymond.

Twenty-three-year-old Raymond Allen Borskey was dead, there was absolutely no doubt about that. He was laid out next to his truck. His truck was parked at the foot of his trailer. Borskey's brain, along with half of his skull, had literally been blown away by a close-range shotgun blast. An open beer bottle lay next to his hand.

There was no real mystery here as to the cause and manner of death. The case appeared to be open and shut. Not much to examine here, either. We would do a full autopsy in the morning. But I still needed to know what happened.

A detective filled in the blanks for me. Borskey had been killed by his neighbor in the trailer park, a thirty-four-year-old man by the name of Mark E. Mire. Mire shot him in the back of the head with a shotgun because Borskey had insulted his dog.

"Wait," I had to interject, "you're telling me this guy's brain got blasted halfway across his yard because of a dog? A *dog*? What was the real deal?"

The detective shook his head. "I know. It's nuts, but that's apparently what happened."

Of course, I assumed there had to be a deeper reason behind this death, but apparently the altercation really was over a dog. I expected money, or a woman, or maybe dope—but a dog? "All I can say is that must be one helluva dog. Where is it? I'd like to see a dog worth dying for to defend its honor. This is just nuts."

The detective informed me that he had not seen the dog, and didn't even know where it was.

So, to recap, the version I got of the story went like this: Borskey was drinking at a local lounge when Mire showed up with his dog. Borskey commented that the dog was ugly. A bartender then asked Mire to take the animal out of the barroom.

Mire went home to the trailer park, got his shotgun, hid in a van, and waited. When Borskey and his girlfriend arrived, Mire stepped out of the van with the weapon. Borskey never saw it coming. That's it. That's the whole story.

Having been raised in north Louisiana, which is redneck country, I had heard that some folks could be a little touchy when it came to their dog, but this was ridiculous.

Mire—dog lover and now "cold-blooded murderer"—surrendered without a battle and was taken into custody. One dead and one going to jail for a long, long time—all because of an insult to a dog.

I imagine alcohol played a major role in this homicide. It does in a lot of them. In the end, all of the death and suffering caused on this night was for naught. I wondered if Mire would try to spin his actions as "justifiable homicide" when he woke up in the morning.

The motive in this case is almost incomprehensible to me. The trivial things that people kill over are astounding. I do not understand it. I did have a moment of clarity the next evening when my eleven-year-old son was sitting with us as we watched the TV news coverage of Borskey, Mire, and the dog. Michael looked at me and said: "That's just plain dumb!"

"Out of the mouths of babes . . ." I muttered.

His attorney argued, albeit unsuccessfully, that Mire should be acquitted because he was "insane" at the time of the murder.

Mire would have plenty of time to contemplate the whole insane affair. He was found guilty of second-degree murder. That's a mandatory life sentence.

So what happened to the dog? The chaos must have confused the poor animal. Following Mire's arrest, his dog ran out on Plank Road as Mire was being led across the highway and was run over and killed.

"WE'LL BE SEEING YOU . . ."

That crime scenes are typically rich in banter should come as no surprise. One of the more common ones in the coroner's armamentarium is: "If you die stupid, we're going to laugh at you!" Another in the Top Ten is: "We've already got the paperwork done on you . . . how you feeling?" Then there's: "If I see you tonight, you won't be seeing me."

Sometimes these shockingly witty quips prove prophetic and may even become legend, in the urban sense. One such quip got its start at the Greater Baton Rouge Gun Show.

Just in case you are not familiar with a Louisiana gun show, I'll take a few lines to describe the phenomenon. The show features about 200 tables staffed by gun dealers. There are all types of firearms for sale to qualified individuals. The tables are arranged in rows, and the attendees wander down the rows in search of deals and treasures. In order to purchase a firearm, you must pass the "instant background check," the state's threshold to make sure you are not a felon.

I have been around guns all of my life. In my younger years

I even explored the possibility of becoming a gunsmith. But my father had other ideas, and that one never came to fruition. I did become a sort of amateur gunsmith, though, and subsequently have become rather knowledgeable about firearms. That's a positive thing, since that knowledge is essential to my profession. Indeed, in addition to my service weapon, I have a collection of various types of guns that hold special interest for me.

In order to appreciate or even understand this story, you will need a little background information. It centers around a particular make of semiautomatic pistol, a gun made by Hi-Point Firearms of Dayton, Ohio.

My first dealing with this particular make of weapon came during our process of cataloguing guns used in suicides. Suicide weapons are held at the coroner's office. As they are not considered murder weapons per se, these guns are the property of the estate of the deceased. But it's rare that family members want them back, and therefore we end up housing a lot of them forever. One of the pistols we cataloged was a 9mm semiautomatic pistol that I will simply call a "Lo End" firearm. It's sort of a "gangsta" Saturday-night special, in my book. But even I will admit that it gets the job done. In attempting to check the chamber of this particular weapon, I noted that the safety was bent, blocking the slide from moving backward. Upon bending it back with a screwdriver and a little pressure, I was able to check the gun. We laughed as we compared the quality of that weapon to our Glocks. We catalogued it, put it in the vault, and gave little thought to it after that.

The "Lo End 9mm" didn't cross my mind again until I went to the semiannual gun show. I was surprised to find several of them offered for sale. The asking price was a little over a hundred dollars. By comparison, what I considered a decent pistol was going for $400 or more at the time. One of the vendors told us that it was a great backup gun, especially if you're a hunter. From then on we referred to it as a "hunting weapon" that every sportsman should have. It became a running joke.

About six months later, I was wandering through another Baton Rouge gun show, checking out the latest guns and hoping to buy ammo for myself at a good price. I was browsing among the two hundred or so gun-dealer tables when I came upon a police officer I knew. He was amusing himself by watching a Lo End 9mm being touted and sold. The gun dealer had a table full of what we consider to be "off-brand" and "cheap" firearms. Some resembled assault weapons. We pretty much know the type of person who is attracted to this type of table, and it is never a true firearm aficionado. The buyer in question, a black male dressed in a rather nice patterned shirt, was listening intently. The shirt is really what caught the officer's eye as it showed various Harley-Davidson motorcycles on a blue background. Both he and I ride Harleys, so the shirt struck us as rather interesting. We stood there and listened as the dealer expounded on the outstanding qualities of the firearm. "Best gun on this table, for the money." We were amused by the sideshow presentation the dealer had mastered. He reminded me of a barker outside a Bourbon Street strip club. I could hear him doing his spiel in the same voice— "Come on in, buddy. Most beautiful ladies in town . . ."

Ultimately the guy buys the gun. His gold teeth glistened as he smiled over his new acquisition. He looked up, made eye contact with us, and my policeman friend said to him, "Hey, buddy, good gun. I'd put a lot of WD-40 on it and the bullets to make it shoot better."

The guy thanked him. The next pitch from the dealer was the type of 9mm ammunition worked best in the gun. He just happened to have some there on the table. We went on about our business and laughed about the Lo End being a hunting gun. The officer mumbled something to me about how the point of his WD-40 recommendation was to "gum up the works and maybe save some cop's life out there."

As we were leaving the show, we ran into the buyer again and said, "See you later, buddy."

Buddy responded with: "Yeah—cool, man. See you."

My friend turned to me and finished his thoughts on the matter, "But he *won't* see us, because he's going to get killed out there with his sportsman's Lo End. . . . Nice shirt, though. I like it."

My response was typical: "You're profiling, but, you may be right." After all, this guy had been a police officer for many years.

We then discussed the issue of "Saturday-night specials" and the "gangsta" market that such a firearm would appeal to. We also talked about profiling and police work.

His whole demeanor changed when we touched on the topic. I guess he had developed a clear perspective from his police experiences. "If I see a carload of white males riding around in an all-black neighborhood at eleven P.M. and I think they don't belong there and are up to no good, am I profiling? Sure I am. I'm *assessing risk* based on what I know. It's not about racism, it's about actuarial data, just like the insurance companies do to assess risk." Then he closed with his final point: We both opined that the person buying that type of weapon from that type of dealer is at high risk for shooting someone or getting shot.

I concurred, and encouraged him to lighten up. "Hell, I'm on *your* side."

At a death scene soon after, we were gathered around a customized two-door Buick. It was about five A.M. and the sky was starting to turn a light blue as it readied for daybreak. The humidity already hung heavily in the air and draped about us. The car was light green. The paint job was immaculate. I guessed it to be a late-1960s model, restored to perfection. The "gangsta" whitewalls and the "spinner" hubcaps reminded me of the old muscle-car days.

It was a nice ride, very clean, except for the blood on the white leather upholstery.

An examination of the crime scene revealed the driver of the Buick had lost control at the intersection, which was about a hundred yards away from the current location of the car. The vehicle had jumped the curb and come to a stop in an empty lot next to a small warehouse. There was no indication that the driver had applied the brakes. The car had rolled to a stop but was still in gear. Evidently the engine died when the Buick bounded over the curb. The driver had a Tootsie Pop in his mouth and a bullet in his heart. He was wearing a rather nice shirt patterned with Harley-Davidson motorcycles on a blue background. My friend recognized him immediately and made a rather concise comment to me on the situation: "Profiling, huh?"

His Lo End pistol was not with him. We surmised that another "sportsman" had taken it. The whole episode seemed surreal.

Coincidence? Educated guess? Who knows? Bottom line is that if you live by the sword, or the 9mm . . . *there's a good chance we'll be seeing you but you won't see us.*

Since I come from a clinical background, there are times when I drift into a clinical assessment paradigm. What that translates into is that I focus on the biological or physical aspects of the corpse first and foremost. But, I also try to consider any signs or clues that might give me some indication of the victim's psychology as well as that of the perpetrator. For example, if the victim has track marks, the possibility that he is a drug addict goes to the top of the list. But he could also be a recent plasma or blood donor or even be a dialysis patient. An indication is not a fact. If he has an appointment card for a visit to a mental health clinic, that's a clue. Or multiple "jail house" tattoos might indicate a multiple offender. If he is a blatant homosexual, there is the possibility that this is a hate crime and the perpetrator has a character disorder. If he has the address of a married woman in his wallet, this might be a passion murder. The list of possibilities may seem endless, but it is important to keep an open mind. Then there is the social context. The type of lifestyle of the victim may well point to the type of person who would kill him. The most obvious being a gang war. And we have to consider the spiritual context, especially with serial killers who might be on a "mission from God." Maybe the killer feels justified if not morally compelled to kill a sexual predator who has exploited his child.

Of course this process of trying to fill in those biological, psychological, sociological, and spiritual blanks can lead to a great deal of speculation with some infusion of my own bias. But, like speculation, this approach can offer some benefit since it does get me thinking in an orderly and comprehensive fashion. I just have to keep in mind how speculative it is. A supposition is definitely not a fact but it can help lead me to a fact. It can be dangerously seductive, too. The danger is that I may try to make the facts fit a "pet" scenario.

Sally Givens is what we'll call her, a single mom raising two kids and struggling to stay afloat financially. It was about four A.M. in the fall of 1999 when she turned her car down a side street off of North Acadian Thruway. She was delivering newspapers before the city awakened. The side street was dark. The streetlight was overgrown with vegetation.

Her headlights revealed a black female lying in the driveway of one of her paper customers. Sally was too street-smart to stop and check it out. She turned the vehicle around immediately and drove straight to the closest police station. EMS was dispatched and arrived in three minutes and fifty-seven seconds. About three hours too late for the victim.

A local who was staggering about that morning informed EMS that he recognized the deceased as a working girl by the name of Tasha. That was all he knew, except that he had seen her in the area earlier, but couldn't remember when. He was quite inebriated. He did stick around to talk to the police, but they got little more from him.

The shrill ring of the cell phone ended my sleep at five A.M. The message awakened me immediately. "Got a black hooker down. Could be our boy is at it again." "Our boy" referred to a person killing black prostitutes. He was also becoming known as the "non-serial killer," a title that originated as a sarcastic retort to those who dismissed the string of black prostitute murders as just coincidental.

De, awakened by the call, asked if she could come with me. "Sure, De, you're legal." *Besides, this may be him and we can always use a woman's touch.*

DeAnn, whom I first met in a psychiatric hospital in 1985, is a medicolegal death investigator and a registered psychiatric nurse. She was obviously grateful that I was allowing her to come, and her response was one of understanding and gratitude: "Always a smart-ass! Okay, just for that you're taking me to Frank's for breakfast after this."

The scene was a typical southern Louisiana morning. The sun was on the rise and the dew was burning off in the form of steam. Mosquitoes were swarming about and the air was heavy with insect repellent. The responders had sprayed themselves liberally with the stuff. I guess they believe in West Nile virus. We followed suit and sprayed ourselves. The usual cast of characters was present: detectives, crime-scene officers, uniformed officers, and gawkers. Amazingly, the press had not arrived yet.

The crime scene was on Randall Street, which runs off of Plank Road. The beginning of Randall was blocked by the crime-scene van. The scene was cordoned off down Randall until the street intersected with another. That end was blocked by a marked police unit, so we had control of the street for a whole block. As we waited to be cleared by crime-scene officers to enter the area, we chatted among ourselves—mostly about the possibility that this prostitute could be part of a string of murders that many felt was the work of a serial killer.

We were cleared to approach the body.

We found a blue purse several feet away from the body. It was open and its contents spilled out over the driveway. Condoms and several twenty-dollar bills were scattered about. The presence of the twenties suggested that the crime scene had not been tampered with to any extent. Nobody in this neighborhood is going to walk past a twenty-dollar bill, let alone six of them.

There was also a six-inch kitchen knife in view. It did not appear to have any blood on it. *Murder weapon? Victim's weapon?*

Tasha (not the victim's real name) was lying face down in the first driveway on the left. The ants had found the body almost immediately. The blow flies were still napping and there was no other insect activity on the corpse. The victim was dressed in blue slacks that appeared so tight as to make the gold-colored belt superfluous. She had on a pink striped blouse and blue shoes. Her hair was shoulder length, but it looked like a wig.

As it happened, my initial speculation was that this was the work of the serial killer. "Wonder if it's him . . ." I mumbled to myself. I mumbled a little too loudly.

Upon which my wife blurted out, "Nope, it's not him." She said this in a very authoritative manner, and of course everyone heard her say it.

Now, I like to think I can put professional pride aside and be open to input from any source. But I must admit this irked me just a little. "Now, how in the hell do you now that, De? How do you know this is not the serial killer's work? We haven't even examined this woman yet!"

"Because I know this 'woman' is a male. I don't even have to see the face. Look at the hands—too big. Look at the feet—too big. Plus the butt is too angular. The body fat isn't in the right place, and besides . . . that outfit is atrocious."

I couldn't let it go. "We'll see." De helped me turn Tasha over. Examination revealed Tasha had on a black sports bra. Each cup was stuffed with a rolled-up stocking. A key was in the right cup and a driver's license in the left one. Tasha was a man I'll call Alton. Alton also had on thong underwear that allowed him to fold his penis back toward his buttocks. De was right. Alton, alias Tasha, was definitely a male.

Still, I couldn't capitulate without a final stand. "So maybe the serial killer made the same mistake—I mean assumption."

De: "He hasn't shot anyone to date, has he?"

Smart-ass. I'd kept that part to myself.

Examination revealed that Tasha had sustained a gunshot wound to his right upper back. The bullet nicked a major pulmonary vessel and caused him to bleed to death. He also sustained a gunshot wound to his left hand. He was able to run from his corner of business to the driveway before he lost too much blood to continue.

I asked De to interpret the blood trail. "Well, he was running from the corner of Plank Road. You can see from how the drops splattered that he was moving fast at first; then you can see where he slowed down. Looks like he may have just stood here for a moment before he collapsed because the drops here are almost at ninety degrees."

One of the crime-scene officers felt the need to share that he was impressed with her interpretation. I had to agree. I was proud of her.

The empty casings were .380 caliber. Some were under the PD's crime-scene van—they had parked right on top of them.

Of course, as the case unfolded, the prime suspect became the victim's lover. That's always the first suspect, assuming he had a lover. There are several lessons here. Even though I know better, I found myself giving in to that seductress known as speculation, who tempts you to look for ways to make the facts fit your preconceived assumptions. That's an investigative pitfall, and one—as Tasha and DeAnn reminded me—I'm not immune to. We all need reality checks now and then. I was jumping to a conclusion. Indeed, on our way to the scene, I had been replaying the other similar victims in my mind. Maybe I had allowed myself to be biased by the initial call. But, I was on the wrong page. The biology was wrong here. The victim was a male, albeit a female impersonator, and one who fooled me at first glance. But he was still a male, and not our serial killer's type. The cause of death, a gunshot wound, didn't fit either. The "non-serial killer" likes close contact, and strangulation is close and personal, so the psychology was just wrong. "Our boy" would never have just shot someone and run off.

There was no positioning of the corpse to display his work, and like I said, bullets just aren't as intimate as choking a person to death with your hands.

The other lesson here is that men and women tend to see things differently. I strongly believe that some subtleties, especially in female homicides, are best detected by a female. A female investigator is a valuable resource to have on your team. Actually, this was only a preview of how that value would express itself later in other cases. Ironically, I would be criticized by my more vocal detractors for allowing my wife to volunteer her services. There were wrong about lots of other things too.

By the way, just as an addendum, Frank's is a local restaurant where lots of us judicial types meet up. We did go to Frank's for breakfast that morning. Our waitress recognized us, and when De ordered crow for me, both she and the waitress laughed. Personally, I failed to see the humor in that. Where's my damn biscuits and alligator sausage?

STAND IN MY SHOES

There is a "switch" that most emergency responders have developed. I've seen it in EMS paramedics, police officers, healthcare providers, and even in myself. This objectivity switch is flipped on immediately when we get a call-out. We try to put our emotions aside in order to be able to render care, or do an investigation or an autopsy, free of subjectivity. We are applying scientific methods to the task at hand.

But just because a person is not reacting emotionally to the events he or she is dealing with does not mean the emotions are not coming—because they are. We repress them at the scene. That's a good thing. We need to be able to compartmentalize situations. That ability allows us to function optimally in a crisis situation. But if we do not deal with those emotions after the acute crisis is over, then they will surely deal with us later. Those repressed emotions bubble up to the surface as alcoholism, prescription drug addiction, post-traumatic stress disorder, divorce, and even suicide. We all know comrades who have burned out.

I think that to really understand what all this does to a person and how it expresses itself, you have to stand in the shoes of a responder. I hope this journal entry clarifies some of it.

One warm, mosquito-infested Saturday night in May of 2001, we were out among the usual responders at the death scene of a teenager. I say "we" because DeAnn was with me. It was just before midnight, and the humid "air" clung to us. He was not a big kid—the eighteen-year-old Broadmoor High student looked about fifteen. He had sustained a single gunshot wound to his neck and had bled to death, alone, there on a main thoroughfare in Baton Rouge.

He had tried to run from his assailants. I could tell this by the blood path he left and the way the droplets of his blood patterned on the pavement. He seemed to be running in the direction of his home. He had a lacerated jugular vein. If he had known to put pressure on it he might have lived. I thought of how scared he must have been and how his death was pointless. *Death doesn't play fair! Death doesn't "play" at all!*

So there he was, crumpled up on the side of the road. His

name was Demetrius White. He was walking home from a nearby apartment complex where he was visiting friends. The police were already on the trail of the killers. Minutes before midnight, three men in a four-door maroon Oldsmobile approached White and tried to rob him. White struggled and was shot in the back of the neck. The criminals went on to carjack three other men in just hours.

This was not a whodunit. Our job was to make sure we collected all of the evidence and did everything just right so we could get a conviction, and hopefully some modicum of justice for Demetrius. I realize our legal system is not necessarily compatible with justice at times. I try not to confuse the two.

We were waiting for a detective to arrive before we moved the body. Evidently it was a busy night for them. Having gone as far as we could, we were sort of huddled up around the body. A cloud of gloom hung over us and no one was talking much. We were in a kind of numb holding pattern.

There are four lanes of traffic on this street and the lanes are separated by about twenty feet of grass-covered ditch. As DeAnn looked across the road she spied an armadillo, a creature common to Louisiana. She was concerned that the creature might find its way over to us. Several of us assured her that it would probably be roadkill shortly, as it was wandering out into traffic.

What the armadillo did was serve as a sort of reprieve or temporary diversion from the depressing task at hand. One of the crime-scene officers expressed concern about the health of the creature. Another joined in to assure everyone that it would be okay. Another was sure it would be dead within ten minutes and offered to start a betting pool on the time of death.

When the armadillo did meet its abrupt demise, several of the group were truly upset and showed it.

We have a murdered child at our feet and we are upset about a damn armadillo. *Are we so callused that we feel more for an armadillo than for a child? Have we lost all perspective?* The answer is a resounding no, though sometimes it may seem that way on the surface or to someone looking in from the outside.

This is a vivid example of the dichotomy that many responders live with. It is a classic example of displaced emotions. We had seen far too much death that night. We had a professional obligation to Demetrius to do the best job possible and that meant staying objective. These were not two totally separate responses. This was a projection of our anger, outrage, and sadness that we felt for Demetrius. These were the emotions we could not talk about or express at the scene lest they shade our professionalism and our abilities to do right by him. The bottom line is that we just couldn't stand to see death win again.

As Elvis sang,

Before you abuse, criticize and accuse,
Then walk a mile in my shoes.

Two years later, on March 26, 2003, a jury convicted a twenty-three-year-old man in the shooting of Demetrius White, committed during a month-long spree of robberies and thefts. Travis Grevious was found guilty of second-degree murder, which carries a mandatory life sentence. The jury also found him guilty of conspiracy, armed robbery, and theft of a firearm. After Grevious robbed a convenience store with help, he stole a man's car from the Nicholson Apartments on the LSU campus. The car was a maroon 1991 Oldsmobile, the same car used in the killing of Demetrius White. Police found Grevious's fingerprints on the car, and the murder weapon in his grandmother's yard. One of the accomplices received ten years for conspiracy to commit armed robbery, and the other man received eight for the same count.

Lonnie Poydras, who admitted being involved in the attempted robbery, testified in court that they had picked the teenager randomly. He quoted Grievious as saying, "There goes a victim."

THE WALK-THROUGH

As I drove down the street toward the crime scene, I admired the arbors that shaded the street and sidewalks. It was spring in Baton Rouge and the vegetation was lush and brilliantly green. It is a time of rebirth. This was an established neighborhood. It looked like something out of *Leave It to Beaver.* The only thing that betrayed the serenity that quiet breezy day of April 4, 2000, was the crime-scene tape stretched about the yard.

Any feelings of tranquillity were shattered by the revelation of the secrets that the house held within. The first thing I noticed as I entered the scene was a male dead on the floor of the carport. I was informed that we had a total of four dead. The other three were in the house. The detective offered to give me a walk-through. His voice was flat, which meant something really nasty lay inside. He escorted me unceremoniously through the front door.

The usual chit-chat was not going on. The place was really quiet—I mean, no one was talking. The objective and speculative comments were at a minimum. Nobody wanted to hear the bullshit about "How when I was in Vietnam I saw holes blown through slopes like that" or "Damn, bet that hurt" or all the rest of the crap that can go on. People tend to project and when a kid is killed, you see your own kid there. No side conversations about last week's fishing trip or agency politics. Flip comments were not appreciated and drew dark looks.

There were toys in the living room. Something that always gives me that chill up my spine is entering a homicide house and seeing kid's toys lying about. The immediate foreboding is that a child has been murdered. Regrettably, the feeling was accurate this time. A young woman's body was in the hallway. I paused and did an initial cursory examination. She had been shot to death. We then entered a bedroom. Another young woman was dead in the bed. Her injuries were compatible with a close-range shot from a high-powered rifle. Empty cartridge cases scattered about indicated the killer was shooting either an SKS or an AK-47 assault rifle.

Then there was the child. The little boy was huddled in the corner of the bedroom. He was four years old and so small and vulnerable. As I went over to him I realized that half of his shoulder was missing and a brutal gaping wound exposed his internal organs.

My thoughts tried to grasp how this atrocity could occur. I

thought of how terrified this little boy must have been. And I asked to no one in particular, "Why does this happen to a child?"

Actually, I probably knew the answer. I suspected the boy knew who did this, and he would have been able to identify the killer.

How is the question I can't get past. I cannot answer it. How does a man point an assault rifle at a child, look into those innocent, terrified eyes, and pull the trigger? It escapes me.

When I lifted him up I could appreciate just how light he was. His arm was barely attached by a few shreds of skin. I tucked what was left of his little arm in front of him. Then I gently put his little body into a body bag. It looked too big for him. I had the crazy thought that he would feel lonely in there. You're supposed to hug four-year-olds and put them to bed, not into body bags.

There was also a toddler in the house, a two-year-old. He alone was not harmed. Indeed, when a coworker of one of the murdered women called the house, the two-year-old answered the phone. He said his mother had fallen. The coworker went to check on them and found the murdered man in the carport, and that is how the crime was initially discovered. The toddler had spent twelve hours amid those dead bodies.

An emergency medical responder who carried the two-year-old from the house said the child told him, "My daddy made my mama fall."

Once in the EMS unit, the child began to call for his mama.

The man charged with this contemptible crime turned out to be the father of the two-year-old toddler. The killer was estranged from the child's mother. She was one of the women whom he murdered there in the house. The other woman was her sister. The child found huddled in the corner was the sister's son. As for the poor guy in the carport, he just picked the wrong time to come a calling on his girlfriend.

The empty cartridge cases found at the scene linked him to the crime. He had several unfired cartridges at his residence, and imprints on those live rounds matched the imprints on the empty cases at the crime scene. They had been cycled—chambered but not fired—through the same weapon. This evidence served to link him to the murder weapon.

That neighborhood and that street are tainted for me now. I know when I turn onto it I think of that gruesome quadruple homicide. A homicide changes lots of things forever. Indeed, the house in which someone is murdered tends to become a sort of morbid landmark. The neighbors are also changed. It's a quantum leap from reading about a homicide in the paper to knowing that it happened next door to you. It shakes the very foundation of the neighborhood. It strikes fear into everyone. That fear tends to linger on, even after the perpetrator has been apprehended.

In the end, the killer was convicted on four counts of first-degree murder. The jury deadlocked during the penalty phase of the trial and he escaped the death sentence and is serving a life term in prison.

THROWAWAYS

A police officer was once explaining to me that there are certain people who are just "throwaways." Nobody misses them, he went on to tell me, and nobody cares if they drop off the face of the earth.

The underlying message here is that if this case isn't solved as a "gimme," it probably won't be solved at all. A "gimme" is when somebody cops to the crime or someone rats the killer out. In other words, no real detective work is really going to take place. No family member is going to be asking about the investigation—not for long, anyway, and it will just go away.

It was another spectacular fall day in Baton Rouge—clear blue sky, temperature in the mid-seventies, mild breeze—just perfect. It was October 31, 2000, a great day to get out and bicycle or walk the dog. And here I was, stooped over a decomposing body, collecting maggots that were greedily feeding upon the rancid flesh. A bird was chirping in the trees above us. I looked up and thought, *What's wrong with this picture?* Then I got back to work.

The smell is one of the worst things about working on "decomps." It is natural to be repulsed by such a thing. But, I reached into my mental bag of pearls. I have collected those pearls of wisdom throughout my lifetime, not just in my professional career. They are quips that others have given me. They are coping tools offered in the form of advice. This one had come to me from a very special friend.

Just think of the person as a person. Think of how embarrassed and ashamed they must feel being seen this way and having these things done to them. It's not their fault. Be compassionate with them.

It works. It allows me to get past the stench and the maggots and get to the person—the person for whom I am now responsible.

We were behind the levee near a tree line of willows that offered little shade because of the position of the sun—it was about noon. The Mississippi River was about seventy-five yards away. The odor was horrific. It seemed as though everywhere I moved, the wind followed and assaulted me with the stench of decaying human flesh.

Think of the body as a person.

People often ask, "How do you do it?" Well, this is one of those times when I just have to get tough and go with it. I keep perspective. *This is a human and she will be examined and handled in a respectful manner.* I think past the current condition of the remains and I envision this woman lying before me as "someone's daughter, maybe someone's mother, maybe someone's wife." Right now she was Jane Doe and she had things to tell me.

A man who had decided it was a fine day to exercise his Labrador retriever on the levee had discovered this poor woman at about ten A.M. Actually, the dog discovered her. The man and his canine companion were walking along the top of the levee. The dog evidently caught the scent of decaying flesh and bounded down the grassy levee toward her. He found the origin of the scent—much to the horror of the dog's owner, who called the animal away and rushed back home to call 911.

The victim had been down at least a week or more. The maggots would give me a more precise number of days. They were my informants. So I collected them from the various body areas, took temperatures of the maggot mass, and netted the flies that buzzed above her remains.

The whole process of fly development occurs on a fairly rigid schedule: we know how many hours or days it takes for an egg to hatch and go through its various maggot stages. There are a few variables that must be factored into the process, but if the flies can get to the corpse, they deposit eggs almost immediately after the person dies. That's when the "clock" starts. As a general rule, I take samples of eggs and all the different maggot stages I can discern.

I then consult a forensic entomologist to identify all these stages. In order to control for the species we were dealing with here, she would also raise some of the maggots into adults. But this time I was lucky, and I had a forensic entomologist right there on site. She did the specimen collection herself and would present me with a report of her findings in a few days.

Most of the maggot activity was at the usual sites—every body orifice. I looked for unusual sites of maggot accumulation, because they also get very active in areas of open wounds. Such accumulations can help lead to the cause of death. That was not the case here. The bug infestation was very heavy. Her skin appeared to move or ripple at times because maggots had found their way up under it. It was a horrible sight. *I have to keep this in perspective.*

I tried to identify some discerning characteristics that would help with identification. She had no face left and her hair had fallen away. She had some teeth. "That might help with an ID." Her skin had decomposed to the point that made it impossible to tell what race she was. I examined her hands and was encouraged by the fact that there was some drying or early mummification of the fingertips. "Maybe we can get a fingerprint here."

There was no clothing about and no shoes. We searched the area for any personal effects but came up empty. She was probably killed elsewhere and then dumped at this location. She was literally "thrown away."

We ultimately got a print and ran it through the Automated Fingerprint Identification System, or AFIS. She had an arrest record. She turned out to be Mary Nevils, a white female, age thirty-eight. She'd had a rough life and was estranged from her family. I am told that she was pretty much on her own. She was supposedly streetwise, and that put her into a high-risk category for this type of death. In the end, we cremated her as a pauper's case.

She turned out to be an alleged victim of serial killer Sean Vincent Gillis, who confessed to her murder and several others when he was apprehended in April of 2004.

"Throwaways." I've thought about that a lot. It bothers me—a lot. If you accept the premise that it's possible for some victims to be "throwaways," your motivation for justice will be less than it should be. In other words, you'll do a half-ass job.

I was back to "cognitive emotive dissonance" in that my thinking does not match my emotion. I was doing my "analyzing" and "intellectualizing" about throwaways a great deal. I could not embrace the concept.

My wife accuses me of ruminating. I prefer to call it processing. Rumination is chewing on something, swallowing it, and then, when it comes back up, chewing on it some more. It's an endless, nonproductive process. It's what ruminants do; think cows—they chew their cuds. I was also accused of being "grumpy." Can you imagine? At any rate, it's times like these when I definitely need a reality check.

I got my reality check. It was at my youngest son's school. On the night of parents' open house, DeAnn and I were walking down the hall. I was still "processing" when I saw it on a wall. It was the answer to the question that had been nagging at me. The question that I was afraid to acknowledge even to myself.

It was a drawing of a child, with the caption: "God don't make junk!"

My philosophy precisely: *We don't do "throwaways" at this coroner's office!*

NINE

Headhunter

THE STRANGE BUT TRUE STORY
OF EDWIN ROBILLARD

On the afternoon of Monday, January 10, 2000, I got a call that a body had been found on the roof of the Prescott Middle School, and it had been there a while. That familiar feeling of nausea crept into my stomach when I heard the words "middle school." I don't do well with dead kids, especially murdered ones. Then I was informed the victim was an adult—that was something of a relief. It didn't make the crime any less a crime, but it did quell my nausea a bit.

Some crimes just have a surrealistic aura about them and seem to take on a life of their own. I guess that almost mystical notion stems from trying to make sense out of situations that just don't. The murder of Edwin Robillard was an unusual homicide from the "git-go."

Sadly, I knew the area from previous official visits in the vicinity. The last time I was in this neighborhood, a dead man was found just opposite the school. He had been beaten and strangled to death with a phone cord. He'd made a lasting impression on me, and I had no trouble finding the school. I did have a little trouble navigating past the numerous vehicles already surrounding the site. A ladder was set against the front of the school building to allow access to the crime scene, and of course there were the news cameras. A policeman motioned me over to the rickety ladder. With some mild hesitation—and, to the chagrin of the officer, with a steady shaking of the ladder— I started up.

Once the precarious climb was accomplished, I had only to glance over to see that there had recently been a fire at the rear of the school. I commented on the blackened area and was informed that it represented an automobile that had been burned there several days before and had subsequently been hauled off. Nobody had thought to check the roof for dead bodies at the time.

The body of a black male was indeed on the roof. The victim was bound and attempts had been made to burn the corpse. That is never an easy thing to do, and it was a failed attempt. His body had been visited by the ubiquitous flies of the area and there was maggot activity.

I had concerns that attempts at visual ID would be useless. If a body is not badly decomposed, we frequently take a Polaroid of the face and show it to people who might know the person—"people" usually being possible family members. This definitely was no Polaroid moment. Yeah, I know the phrase is "Kodak moment," but out here on the streets we often take a Polaroid of the victim's face if the face is not too distorted. We can then use that photograph to show family members for possible identification on site. A Polaroid is just a tool, it is not used for a definite or final identification of the deceased.

I had no positive ID, I wasn't sure of the cause of death (though I *was* damn sure it wasn't natural), and I don't know the time or even the day of death. In short, I had a mystery on my hands.

A murderer could conceivably go free if I screwed up. And this murder was becoming more and more complicated as events unfolded. Complexity, of course, equates to a greater likelihood of mistakes, I have only one shot at the evidence, and from an evidentiary standpoint, once the body is disturbed, things are forever changed. I approach every homicide with the same professionalism. And therein, as you shall see, lies my salvation.

I have a limited budget and manage to operate within those parameters in part thanks to the good will of volunteers and professionals who have never failed to step up to the plate. Why? Because it is the right thing to do. I called in Mary Manheim, forensic anthropologist at the LSU FACES Lab, and Dr. C. Lamar Meek, the best forensic entomologist I know. (FACES is shorthand for Forensic Anthropology and Computer Enhancement Services.) Dr. Meek died June 27, 2000. He was fifty-six years old. His untimely death was a loss to all of us. I might add that both of these folks were always very generous with their time.

Dr. Meek collected his specimens and carted them off for examination and rearing in his bug chamber in order to be sure of the species. They would tell him about the time of Edwin's death.

Even with the decomposition, we were able to get fingerprints from the corpse. That was a real break, especially since he had prints on file.

Mary (aka "The Bone Lady") and I agreed that she should be involved throughout the autopsy process, as we might need her services not only for dental comparison and identification but also for determination of the cause of death. It's amazing what those bones can tell you—if you know how to ask the right questions, and how to listen for their answers.

Autopsy revealed the victim had sustained head trauma, but there was a confounding variable: he had previously undergone brain surgery, and the surgical intervention had, of course, required the removal of part of his skull. So it was essential that we determine and document the skull damage due to trauma sustained at the time of his death and distinguish it from the healing skull abnormalities resulting from the surgery.

Edwin Robillard's head was disarticulated from the rest of his body. Mary cleaned the skull at her LSU lab, and due in part to her efforts, we had our answer: Edwin Robillard had a fresh skull fracture in the area that also was affected by the surgery. He died of head trauma. We were now ready to defend our findings and allow any challenging expert to examine the evidence.

The skull was subsequently put into evidence and the rest of Edwin Robillard was cremated, as a pauper burial, at the expense of the coroner's office.

Job well done. At least, I thought so. But this opinion was not shared by several disgruntled ex-employees of the coroner's office, or by Mr. Robillard's ex-wife, who sued me on behalf of her sons. I was accused of being a "headhunter."

The suit claimed that I removed forty-one-year-old Edwin Robillard's head and stored it in a forensics lab without their permission, and that I unlawfully returned his remains to his aunt. The sons also said they did not receive formal notification that their father was killed until they saw a news report of the slaying.

Efforts to discredit my actions were amplified when a "news" reporter from a local network showed up in my office with a camera and some baseless and absurd allegations. I felt like I was in the Twilight Zone when I saw the resulting piece on TV. I'm not sure what amazed me more, the misinformation or the disinformation. The reporter was Marvin Hurst of Channel 2. I like to think that Marvin was misled by the lawsuit and was not just a sensationalist reporter seeking to advance himself at the expense of responsible journalism. I was told that another news agency had been approached with the story, but that agency saw through the smoke and mirrors and declined to go with it.

But the piece did beg the question: Why did I keep the skull? Was it so that I could prove the body was Edwin Robillard or was it so that in my macabre, narcissistic way, I could parade it into the courtroom, as one disgruntled ex-employee would later claim?

Of course I never intended to bring the skull into the courtroom—how ludicrous. That was a spiteful and ignorant statement. The skull was being retained so that once the killer was caught, the defense expert and any other experts the prosecution might need to call upon could do their own examinations. I fully expected the cause of death to be challenged in criminal court, and I wasn't going to leave any loopholes for the defense to wriggle through. I would also expect any competent death investigator to know that. Bottom line—no murderer is going to get off because Louis Cataldie failed to retain critical evidence. No way.

When I received notice that I was being sued, the case seemed moot to me and I fully expected a summary dismissal. The Louisiana Constitution, Revised Statue 33:1563:B.(4)(a) clearly states: "[The coroner] may remove and retain for testing or examination any specimens, organs, or other portion of the remains of the deceased that he may deem necessary or advisable as possible evidence before a grand jury or court. . . ." I was also astounded by the fact that the plaintiffs' attorney was Rob Marioneaux, a Louisiana state senator.

Here was my thinking about that—obviously flawed! Legislators are responsible for making the law and changing those laws that need changing, right? So how can a senator sue me for obeying a law that originated in his own state senate? I mean, I am sworn to uphold the laws of our great state. He is the lawmaker! What's the deal here?

But my attorney assured me that in Louisiana a senator can make a law, make me responsible for upholding the law, then sue me for fulfilling my responsibilities under that law. Is it just me or does this sound screwy? I still don't get it. Be that as it may, the case continued.

We went through the deposition process, then the various forms of attorney-to-attorney negotiation, and then finally to trial. On the day of the trial, my attorneys approached me to see if we wanted to make one last offer to the plaintiffs before the trial got under way. My response was simple, direct, and loud: "Hell no!"

On the stand, I was accused of being a head-collector and assisting Mary Manheim with her "collection." And at one point, their attorney asked me what I would tell the sons of Mr. Robillard or how would I explain this atrocity to them. I had no trouble at all looking them straight in the eyes right there in the courtroom and assuring them that their dad was not going to be denied justice by any error on my part. I'm just guessing here, but I don't think that's the response the senator wanted to hear.

I left the courtroom after my testimony. In January 2004, my lawyers called to announce that not only did State District Judge Mike Caldwell rule in my favor, he ordered the plaintiffs to pay court costs. I was relieved but not elated because the whole lawsuit and trial seemed so useless and I'm sorry Edwin's sons had to go through all that. It's sad enough that their father was murdered, but to worsen the trauma with some lame attempt at litigation seemed heartless to me. My code of ethics at the coroner's office was simple: "DO THE RIGHT THING." I can even tolerate mistakes as long as no one tries to cover them up.

Edwin Robillard's skull remains at the LSU Forensic Anthropology Computer Enhancement Laboratory awaiting a suspect. His slaying remains unsolved.

TEN

Monster
on the Loose

KILLER CITY

Baton Rouge is a violent city. Now, that's not the most politically correct thing to say, especially if you want to win political friends and influence same, but it's true, and trying to minimize, ignore, rationalize, or otherwise hide the fact can only make things worse. Most of the murder and mayhem here occurs within a few relatively circumscribed areas, and often these crimes are "impulse" killings, associated with passion, ego, robbery, and/or drugs. Of course there is also the occasional "get-back" killing, in which a person is killed as a form of retaliation or retribution for some past transgression, real or imagined. But if, avoiding these specific areas or situations, you consider yourself relatively safe, insulated from such crimes, you're only fooling yourself. To use the vernacular: "It just ain't so!"

It's the ones you don't see coming that get you. Since I have been in office, Baton Rouge has been plagued by at least four known sets of serial killings. I also suspect we have had other murderers come to town, kill, and move on.

At least thirty to forty killings of women in the Greater Baton Rouge area in the past decade remain unsolved, according to the Baton Rouge Police Department and the East Baton Rouge Parish Sheriff's Office. The victims are young and old, black and white, rich and poor. No one, and I mean *no one,* is immune from being relegated to the role of victim by these malicious, twisted predators.

My first exposure to a true serial killing was in the abandoned Dynasty Lounge, in the 1100 block of North Boulevard in September of 1999.

Though it is not confirmed, it's my firm conviction that this murder was one of an infamous series known locally as "the black prostitute murders." These killings purportedly began in the late 1990s, and many remain to be solved. At least ten women have been killed in a four-mile-square area east of downtown since 1999. Most of the women were black, in their mid-thirties, and were found nude or partially nude in or near parks or vacant buildings. Most of the women had rap sheets and had been arrested for drugs or prostitution. It was in 2000 that a couple of young BRPD detectives, Ike Vavasseur and Keith Bates, just back from FBI school, spearheaded a sort of summit to review these deaths as being possibly the work of one person, a serial killer. Multiple agencies, my office among them, participated in the summit, but the idea of organizing a task force just drifted away. I suppose we could say these bright young detectives were ahead of their times. Later we would see two serial killer task forces; but I am getting ahead of myself.

A suspect, Sean Gillis, had been arrested, and though his confession may account for some of these victims, it doesn't account for all of them—and specifically not for this one. Her name was Florida Edwards.

It was dusk when I arrived at her death scene on that balmy summer night, and as expected in southern Louisiana, balmy rapidly gave way to brutally hot and humid. *What else is new?* I was met by a detective and given a short briefing, which pretty much consisted of "We have a dead woman inside . . . we were notified by an unknown person via a phone call . . . it's hot as hell in there, Doc." Then I was channeled down a well-worn path that ran along the side of the lounge. The whole building was almost completely surrounded by thick vegetation, some of which disturbingly resembled poison ivy. I kept waiting for some rodent to rush out and bite me. *On your toes, Lou . . .*

I was met by another detective at the back, and only, entrance to the place. The rear door was off its hinges and served sort of as a bridge from the trail into the usually dark recesses of the lounge. He essentially repeated what I already knew: there was a dead female inside. He told me he felt a need to go outside for a few moments to cool off. I know this guy, he's tough, and if he needed to go outside to cool off, that was a very, very bad sign.

When I looked into the doorway, my olfactory sensors were assaulted by the stench. Equally harsh was the iridescence of the crime-scene lights emitting a glare from inside the lounge. I turned around to go back to my car to get the necessary supplies, which would include a pair of heavy-duty rubber boots.

Once I had the appropriate foot attire, I began my journey back to the crime scene. That journey took me through a small crowd that had started to gather outside. I know that once I am on the business side of the yellow tape, I am insulated from them at least temporarily. But I was accosted by a reporter prior to making it into that safe haven.

Her question was simple and prefaced with a disclaimer: "I know you can't tell me anything yet. I only have one question. Should I stay out here or go home?"

I paused, thought for a second. *She's a decent person and a decent reporter, so what the hell?*

"You'd probably be upset if you didn't stay for this one."

"Thanks, Doc. When will you know something?"

"That's two questions." *Typical reporter behavior, you'd think I'd get used to it.* "Later," was my rather cryptic reply to question number two.

The lounge was truly one of the filthiest places I have ever had the misfortune to enter in my entire career. Evidently this hole was frequented by the dregs of society, such as drug addicts and prostitutes, the so-called high-risk-lifestyle people. Strewn across the dusty floor were empty liquor bottles, used condoms, used syringes, feces, clothing matted to the floor, and bric-a-brac from a rotting building.

It was dark, dirty, and dangerous. Inside, I was greeted by the acrid odor of urine and feces—nauseating. I shuddered to think what other secretions were present. In addition to that stench, there was the foreboding smell of stagnant mold and mildew. I became very aware of my breathing. *Who wants a lungful of that medley of potential pathogens?*

Crime scene had set up several shop lights against the southern wall of the joint, that being the back wall; and while that helped us see, the lights were generating a tremendous amount of heat. They seemed to increase the temperature beyond human tolerance. No wonder people were taking breaks to go outside to cool off. Cool off in the Baton Rouge heat—that says it all.

The illumination did allow me to survey the area. One of the first things I noted were the falling ceiling tiles. *Look out overhead, Lou.* Surprisingly, there were some bent-up lighting fixtures still dangling from above. They had been passed over by scavengers because they were aluminum and not worth the effort. The same could not be said for any of the copper that may once have been there. Indeed, there were some ragged holes in the walls that indicated someone had "salvaged" the copper piping. The walls were greenish from the mildew stain. I thought of my lungs again and the challenge to my immune system.

The floodlights also illuminated the corpse, which was near the front, or northern, wall. Indeed, the whole immediate area of the crime scene was revealed in stark detail. The shadows of the detectives and crime-scene personnel were magnified against the front wall. Those shadows danced back and forth on the wall like specters involved in some macabre ritual or play—a poorly choreographed one at that.

She had been strangled. She had also been positioned in a

sexually suggestive manner. I studied her face with my flashlight from several angles. I could usually recognize mental patients or addicts that I had committed in psychiatric hospitals. I did not recognize her.

As I continued my examination, I noticed an empty five-gallon bucket standing upright about four feet from her body. I shined a light into it and noted a spiderweb replete with what I thought was a brown recluse. *Do not disturb!*

My protocol or custom in this type of setting is to go through a physical examination of the person. I look for trauma, of course, but other things that may help, too. She had stretch marks on her abdomen, which indicated she had children. This is someone's momma. I look for bruises, bite marks, weapon marks, and areas to swab for DNA.

I start my mental checklist. *Was she strangled from in front or from behind? Does the pattern tell us if her killer was right-handed or left-handed? Did she fight back? Will she have his DNA under her fingernails? Can we get fingerprints off her body? How do we best preserve trace evidence? What was the time of death?*

My back hurt from bending over. It was so hot! We were all sweating. So much so that I cautioned everyone not to let their sweat, and therefore their DNA, drop onto her body. I had to take a break. As I stood up to relieve my back, I noted that there were fresh sweat drops all around her. She was practically outlined in sweat—my sweat. I pulled off my gloves and noted my "granny fingers" that had resulted from all the sweat that had accumulated. *What would Granny say now if she saw me standing here like this?*

When I walked out into the Baton Rouge heat, it actually

did feel cool to me. The detective handed me a bottle of water. We stood there talking about how clear the sky was and how nice it would be to be at the beach now. We seemed to be holding a casual conversation while a horror was just a few feet away from us, waiting for us to return. It's a form of stress decompression. It's a way to clear your mind a little before going back into the battle. It helps.

Once back inside and next to the body, I produced a thermometer—similar to a cooking thermometer—from my equipment case. It has a digital readout button that sits atop an eight-inch stainless-steel shaft that looks like an elongated ice pick.

I noted the ambient temperature in the room and called it out so that everyone could note it, along with the time it was taken: "Ninety degrees Fahrenheit." I then drew a small circle on the skin of the victim just over the liver. I initialed it and made a small incision with a scalpel in the circle. The reason for the circle is that I didn't want anyone to think that she had been stabbed by her killer. I then inserted the thermometer shaft into her liver in order to get a reading of her core body temperature.

I waited for the digital readout to stabilize then called out the temperature and the time: "Core temp is eighty-nine degrees." As a rule of thumb, a body generally cools at about one degree Fahrenheit per hour. That's a very rough estimate and an oversimplification. But the core temperature often tells us something about the time of death. One thing is certain, if you don't get it as soon as possible, you can never go back and get it. Her body was cooler than the room. She had been there *at least* ten hours, probably longer. We had heated the room up with the lamps by at least one or two degrees according to my calculations.

Once that was done, I checked her for rigor mortis. Her whole body was stiff. As we placed her onto a homicide sheet and into the body bag, I checked for livor mortis, which was fixed in place. Indeed she had been here for over twelve hours, and maybe even twenty-four. We would process her more at the morgue. Then I could allow myself to reflect a bit on this murder.

Suffice to say that this crime-scene experience was *different*. I felt like I was in the midst of pure evil. I know that sounds a little crazy, but that's how I felt. I've only had that feeling a few times in my career. It's a feeling I get when I think a scene has been staged or altered, and her body seemed to have been intentionally arranged in a sexually exploitative manner. It's that intuition that something is wrong. That things don't fit. It's a tough thing to get a handle on.

And it causes me to wonder. It's been several years and I still wonder about it. This scene fit the dehumanizing dynamic that some killers show. They turn their victims into objects rather than seeing them as persons. The life, the personhood of the victim has no meaning for them.

Theories as to why someone would kill another human being in this manner, in this place, abounded. *Maybe it was drug related. Maybe it was a revenge thing, a "get-back." Maybe it was a message of some sort. Whatever the reason, this murder scene seemed staged to me.*

I've read every book on serial killers that I could find, including profilers' studies, studies on specific killers, and testimony of experts. I have examined the bodies at scenes and attended the autopsies. The result of all this study and research is that the more I know, the more I realize how much I don't know. And, of course, the books can't portray the feelings associated with these atrocities.

I have also come to realize just how little any of us really knows about these monsters. There's that old saying, "If you're so smart, why ain't you rich?" So I ask myself: "If we know so much, why ain't he caught yet?"

To date, we still have not identified a killer in Florida's murder. We were anticipating and dreading the next victim . . . and the next . . . and the next. . . . As previously noted, Sean Gillis doesn't appear to be responsible for this killing.

What does that mean? Is there another killer still on the move? Did things get too hot for him so he moved on for a while? Did he get picked up for some other crime? Did someone kill him? We didn't know all the answers then and I don't think we know them now. The one big question in everyone's mind was *"Will he be back?"* As you will see later, the grim reminder came in 2004 in the form of a another dead woman with a high-risk lifestyle.

FIRST OF A GHASTLY SERIES

While the murders of black prostitutes that had started in the late 1990s faded from the media and the consciousness of the community, I still felt the killer's presence. Maybe he was dead, incarcerated, or had just moved on, but I felt he was still there, still *here*. Every time I was called to a female death, I wondered if I was going to be presented with the same scene.

Two years later the killings started again. But these were different. Gina Wilson Green, an attractive white forty-one-year-old registered nurse and manager for Home Infusion Network, was found dead in her home on Stanford Avenue in Baton Rouge on September 23, 2001, an apparent victim of strangulation. She was divorced and lived alone in this area referred to locally as Southdowns. A homicide in this upscale neighborhood, not far from Louisiana State University, was a rarity. A coworker had come to the house and discovered the crime. From the preliminary information I'd received, I knew I'd need every resource I could muster.

Accompanying me to the scene was our forensic pathologist, Dr. Michael Cramer, a wiry, meticulous man in his early fifties, and a behavioral psychologist who is a close personal friend of many years, Stan Granberry. Stan's also a deputy coroner and provides his services gratis. Gina Green's crime scene really got to him. My grandma would have said he was "spooked." Stan had been on calls with me before, but I think this was just too close to home for him. One block over is his house—*was,* I should say, since he and his family moved to another location not long after. His wife is an attractive blond career woman, much like Gina. He'd helped me sort out other cases, and he is often my sounding board when something is eating away at me. Just getting one's demons out into the open tends to make them less harmful. And it helps to realize one's limitations and boundaries related to crimes. He has a vast knowledge base on personality disorders, like those affecting sociopaths and sexual deviants. Well, I needed access to that knowledge base right now.

The moment I was escorted by the detective into Gina Green's home, I knew intuitively that we were dealing with a different type of homicide. I still have difficulty describing the sensation I felt upon entering the place. It was as if I had been placed into a scene like one of so many props. It was as if the killer expected me to show up and had prepared for my arrival. Stan felt it also and commented upon it and indicated we'd both need to chat about this later. *"Chat" has a whole different connotation when it comes from a shrink, even if he is your buddy.*

Regardless, it's important for me to try to sort these things out. Several things could have contributed to that impression. Gina's house was similar to those of my friends. Indeed, several of my close friends, and even my oldest son, Christopher, lived in the area. Her home was furnished and decorated tastefully, in what I refer to as traditional Baton Rougean style. The color scheme, fabrics, and furniture are classic Deep South: antique furniture (French provincial), Persian rugs, heavy expensive drapes, wooden floors, subdued pastels on the walls, wooden-framed pictures and paintings. Elegant yet comfortable—not stiff—almost understated. Obligatory chandelier in the dining room, roomy kitchen. So maybe that's partly why I got a feeling of violation. Indeed, as I would discover later, it made several of us feel vulnerable. It scared us.

I remember questioning myself as I walked through her house and felt that sensation. *Is it just me or is it time for a reality check? Reality is that this woman is dead! Murdered!*

I first examined Gina in her bedroom. She was in her bed

and covered up with bedclothes. She looked like she was asleep at first. With me in the room were the crime-scene tech, Stan, Dr. Cramer, the detective, and one of my investigators from the coroner's office. Yet it was so quiet—solemn. The room looked normal. Nothing thrown about. Gina was in bed and actually looked peaceful. Then the final covers were removed and she was revealed to be totally nude and her legs positioned in a crude manner. The scene went from one of serenity to lust murder with the turn of a sheet. This stark change of perception engendered feelings of outrage and anger. The killer had gotten his shock effect.

Every piece of bed clothing was carefully marked for evidence and collected. We went by the numbers in a very structured and orderly manner. We used an alternate light source to detect trace evidence from her nude body. This was one of those times when we were grateful to have a state-of-the-art forensic light source. The light source allows us to see organic matter that is essentially invisible to the naked eye. The principle is relatively simple. If we can illuminate the organic material with the right kind of light, it will give off a reflection with a longer wavelength. In other words, it will fluoresce. We adjust the light source for the right wavelength, put on glasses and can locate the evidence. When I had my first exposure to a forensic light source I was both amazed and horrified. A salesman asked me to go into the men's room with him for a demonstration. I know, it sounded a little weird to me, too, at the time and that's not what horrified me. We went into the bathroom and he showed me all sorts of "invisible" biologicals that I was happily unaware of prior to that demonstration.

That horrified me. To make things worse, I found out that the bathroom had just been cleaned. Now that *really* horrified, but it also amazed me. The conclusion was obvious. A killer could clean up the crime scene to destroy trace evidence but he'd have to be using this light to make sure he got everything. Most killers don't carry a forensic light source around. I was sold on it. Imagine turning out the lights and it's pitch dark. Then a blue hue moves slowly over the body. You see only a blue glow unless you're are wearing special glasses. Suddenly an irregular orange splotch appears on the inner thigh of the victim. It is a semen stain. Your heart almost stops. *There it is—jackpot—we got lucky and he got sloppy. Careful with the evidence, don't want to lose it.* It will be reconstituted with sterile water and placed into evidence. I look at the evidence container and my mind addresses the killer who had set this staging for us. *This is it. This will give us your DNA ID and tie any other of your murders together. And when we find you, it will nail you. And we will find you. Gotcha, asshole!*

That scene is still vividly with me. There we were in this woman's bedroom examining her lifeless body under a technical light source. It was so eerie. We looked like alien scientists with our orange-colored glasses on as we continued to search for any additional stains or fibers that might be present. We went over every inch of her skin. Examination of her neck indicated that she had been strangled to death.

Did she know this guy? Was he waiting for her inside her home? Did he break in? Did he cajole his way in? The back door was unlocked, and there was an old dog in the yard. It didn't even bark at me. Maybe she opened the door to let the dog out and he was waiting for her. She had an alarm system and there was an alarm call at 3:47 A.M. on September 23. She had gone through the house with phone in hand and the alarm company on the line. Evidently she felt secure and that was the end of it. How sadly mistaken she was. *Had the killer been in there with her and she missed him during her house check?* There were signs of a struggle in other areas of the house. Crime scene collected her blue patterned shirt, the one she had been wearing the night before. On it was a nickel-sized drop of blood. It did not belong to the victim. (Later it would prove, through DNA testing, to match the semen.)

The killer took her cell phone, which was later found across town in the Choctaw/Airline area.

The whole chain-of-custody process was carefully followed—from her home, through the autopsy, and to the ultimate release of her body to the funeral home. From an objective standpoint, it was seamless. We did it right. I use the collective "we" for all parties concerned at the time. That's why I am absolutely confident that our evidence will stand up to courtroom scrutiny. If this sounds like pride or arrogance on my part, it is not. On the contrary, it's about humility. It is the simple acknowledgment that I fulfilled my obligations and responsibilities to Gina and her family. It's what I'm supposed to do. It's all I can do at this point. It's about justice.

From a subjective standpoint, I've walked through Gina's house and reworked that scene in my mind more times than I care to count. I guess I'm trying to reconstruct events in order to try to understand how someone is capable of such a crime. *What was he thinking? How can you so dehumanize someone?*

Did he stalk her? What caused him to choose her as the victim? How could this have been prevented? Was there something there that heralded a subsequent murder? Even though we know who her killer is now, there were still so many unanswered questions then. I don't know if those questions will ever be answered. Until then, I'll continue to replay that tape in my mind—frame by frame.

Autopsy confirmed that she died from asphyxiation due to strangulation. Evidence showed that she had also been raped. DNA later linked her to Derrick Todd Lee. He's been confronted with immutable evidence, linked by DNA to seven female murders, and already convicted of two of those murders. Yet ruthless, uncaring, unremorseful sociopath that he is, he refuses to fess up to his other actions and give those surviving family members any relief or closure. Like Stan says, "What do you expect, Lou? He's a sociopath."

GERALYN

It happened on January 14, 2002, in Addis, Louisiana. That's in West Baton Rouge Parish, so it was out of my jurisdiction and not my case. I wish it had been. Things might have been different. Geralyn DeSoto, an attractive twenty-one-year-old white female, was beaten and stabbed to death in her home. Her throat was slashed. She was a graduate student who had registered at LSU on the day of her murder. Her cell phone was missing. Her murder was not tied to Gina Green's. I can't be sure what was initially done or not done at autopsy. It was not

my case. I did, however, talk to a family member later who insisted that Geralyn was the victim of the Baton Rouge Serial
Killer and that none of the various law-enforcement officials
she spoke to gave her a serious hearing. Some questions arose
about the handling of evidence and about whether a rape kit
had been used, but nothing came of it. There is also a gaping
question about the rape kit. Was one even done? Again, Geralyn's death did not come under the jurisdiction of my office
and subsequently we had no involvement in that autopsy.

Months later, her fingernails, samples of which were taken
during autopsy, would prove her family right. The killer's DNA
was found there. He had not gotten away unscathed, and the
details of her final moments of life would be reconstructed and
presented in graphic detail during trial.

MISSING

The rains in Louisiana can get pretty heavy at times. They can
wreck homes, ruin crops in the fields, flood roads, and occa-
sionally help find a body.

One of the greatest unsolved mysteries in my career brought
me to a levee of the Mississippi River on a rainy Father's Day in
2002. Earlier that Sunday, two retired gentlemen from the area
had their hearts set on fishing in a bar pit behind the levee that
keeps the Mississippi from flooding East Baton Rouge Parish
and all points north and south.

A bar pit is created by the dirt taken from the river side to
build the levee. The pit fills with water and is restocked with

fish every time the river rises out of its natural banks. The original term, "borrow pit," over time became "bar pit." But today, due to rainy weather that made "gumbo mud," the two fishermen predicted the bar pit would not yield the best catch.

As I understand it, they looked around and spied a pond on the far side of the levee. Even better, they saw the remnants of an old dirt road that would allow them to drive up to the pond. This road was essentially two ruts that ran by the side of Ebenezer Baptist Church, and they followed the two ruts until they reached their destination.

They set up on the litter-strewn banks of the pond, fished for a while, then concluded this was just not a good Sunday to be fishing. A joint decision was made to abandon fishing and seek other endeavors. So they loaded up the buckets and poles and the ice chest and set out to retrace their way back to River Road. The rough road required them to move very slowly, almost at a crawl. One of the luckless fishermen was practically hanging out of the window when he spotted something strange in the road. When he got out and kicked it out of the way, he suddenly realized it was a human skull.

I was with my wife, DeAnn, at home when I got the call that a skull had been found on River Road. As we drove south along the curvy blacktopped road, DeAnn and I talked about the case. As a psychiatric nurse and a registered medicolegal death investigator, she has the training and experience to investigate deaths and collect evidence for a coroner or medical examiner. Her unique combination of skills has afforded us valuable insight into many homicides, particularly those involving women.

When we pulled up to the picturesque little wooden church, the rain continued, and any resemblance to a Norman Rockwell scene was abruptly dashed by the detectives from the East Baton Rouge Parish Sheriff's Office who greeted us. We knew each other from previous homicides and enjoyed a good working relationship—not always the case between agencies. Their crime-scene officer was a seasoned veteran in death investigation. De and I were motioned over to a canvas shelter that had been erected over the skull. The skull was about twenty feet or so from the edge of River Road—the crime-scene officer of course had taken more exact measurements.

Initially I examined the skull in place. Once I was assured that all the necessary photos had been taken, I picked it up for a closer look. At fifty-four years of age, I'd already taken part in innumerable autopsies and handled innumerable specimens, but the eerie sensation of holding the skull of a human evades description.

It was as though this person was looking back at me through those hollowed eye sockets and trying to tell me what happened. My job is to respectfully examine and analyze every possible facet of information available. It is here that my duties as victim's advocate, conservator of the peace, and physician all merge. It is here that I must call upon all my expertise and resources to "listen" to what the victim has to tell me. It's a heavy responsibility, and this ain't no TV moment. This is for real. I continued my examination.

There was still some tissue inside the cranium. Several teeth were missing, as was the mandible, or lower jawbone. The bone structure was comparatively delicate and the size of the skull was such that I surmised these were the remains of a female. There were enough teeth in place for dental comparison and we could surely make a DNA match if we had some DNA to compare it to.

I instructed one of my investigators to move out and search for other bones. DeAnn and I did the same. A light rain was still falling, and the world seemed to have become a hazy gray— a very depressing gray. It is amazing what people do not see when they are not looking for anything in particular. Scattered about the churchyard were the skeletonized remains of a young female. Scattered skeletal remains that showed evidence of extensive animal activity. We soon learned that a stray dog had taken up residence there in the past several weeks.

Most of the ends of the long bones were chewed off. The dog had been bringing this woman's bones from her murder site up into the yard and eating them! Members of the church had been parking in the grass yard on top of them. They had done so as recently as this morning. No one ever noticed. It is times like this that make me think the whole world is insane, or blind, or both. The preacher later told detectives that he had actually thrown some bones he found in the yard into a nearby ditch.

We began a systematic search of the area. We found ribs in one area and vertebral bones from her back in another. There were parts of her pelvis scattered about with the bones from her arms, legs, and hands. The bones were strewn over an area that would easily accommodate twenty to thirty parked cars. And, to complicate matters even more, there were also animal bones— mostly chicken and turtle bones—in various places around the church. Poor Phideaux (or so we had dubbed him) was taken into custody by animal control. I doubt there was a happy outcome for him.

If I ever had a need for a forensic anthropologist, it was now. I called Mary Manheim and laid it out. As usual, she responded to the situation right away. She would grid it all out and process the remains in her lab.

By now, we all had a high index of suspicion that we were dealing with the remains of Christine Moore. At least, it was more probable than not.

I had read about Christine Moore in the newspaper. She had gone missing some three weeks earlier, on May 23, 2002. Her car had been found ten miles away from this area, at Farr Park. A five-foot, one-inch twenty-three-year-old black female, she was in graduate school at LSU, studying to be a social worker. Her weight was reported as 115 to 130 pounds. She was a jogger. She was the oldest of seven children and the valedictorian of her class at Xavier Preparatory School. Her father called her a trailblazer.

How can a young girl's bones be strewn about a churchyard and nobody notices it? Is it just me? Am I the crazy one? Actually, I found it a bit reassuring that DeAnn was experiencing the same level of disbelief. Of course, there is the phenomenon of shared paranoid disorder, but . . . *Enough! Back to business.*

We needed to find the primary site where the body had been dumped and where it had subsequently decomposed. Unfortunately, our best witness was the dog. We tried to search the area but came up with only a decomposing deer in the field behind the church and a dead dog in a ditch up the road. We decided to call in a dog of our own—a "cadaver dog," one specially trained to seek out human remains.

Christine's body had been out here for about a month. Decomposition is a relatively rapid process in Louisiana. Between insects, carnivores, and the temperate climate, skeletonization can be accomplished within that time frame. Somebody must have smelled the decomposing remains, but they must never have considered the possibility that it was a human that was decomposing. Actually, given the fact that we had come upon the two aforementioned animal carcasses, it was possible that the odor could be mistaken for just another animal. Still, it still seemed illogical that no one would investigate it, especially in lieu of the publicity about the missing girl. Did no one around here keep track of current events?

The specialized cadaver dog had arrived. The dog, a German shepherd, lived up to its reputation and soon led his handler to the decomposition site. It was there that another horror scene awaited us.

We followed along a rudimentary path that ran from the rutted road to a small clearing about six feet in diameter, and when we got close, the odor was unmistakable. And the physical evidence further supported the fact that this was where Christine had lain for the past four weeks. Specifically, her hair had fallen off her head here during decomposition, and the ground was discolored from the leakage of body fluids as she decayed and the dog feasted on internal organs.

The rain was letting up now but the humidity was still high and we were all drenched with rain and perspiration. Our hair was matted down. Our clothes were wet and frumpy, and we looked like we had just crawled out of the swamps. I noticed that we had inadvertently formed a sort of odd circle around the spot. No one talked for what seemed like an eternity. The horror of it all was still sinking in. We may have looked like defeated swamp rats but the atrocity before us only made us more determined to get justice for the victim. We systematically began to investigate this part of the crime scene.

Her jogging shoes were to the south. Her hair was to the west. It was black, the color we would expect to find if this was Christine. Several other of her bones were recovered. In the end, we would find about seventy-five percent of her skeleton. Some of her clothing was also recovered.

We trudged back to the churchyard. Thirty or so little yellow marker flags were rippling in the breeze. They fanned out over the green grass and past the picnic tables, reminding me of some sort of field game that children who were members of the church might play on family day. But these flags were not festive in nature; these were morbid in that each marked a bone or bone fragment—evidence to be collected.

The autopsy confirmed what we had suspected: the dead body was Christine Moore's, a positive ID having been made based on dental records. A skull fracture indicated that severe head trauma was the probable cause of death.

I could envision her being preoccupied with whatever she was doing. *Maybe she was just getting out of her car when he surprised her. Maybe she was preparing to clear her mind with a long jog on that river road. In that split second before she became unconscious, she would have known that she was in trouble—*

indeed, from that moment on he was in total control and she was doomed.

Christine's hair was sent to the crime lab for trace evidence analysis. Her father had requested a lock of hair, but we could not comply with his wishes at the time because it was evidence. He understood. I first met Tony Moore, Christine's father, in the morgue, about a week after the autopsy. He came with the funeral home people to retrieve her bones. A local priest was there to bless her remains prior to her journey home. We placed her bones in a pink coffin. I remember thinking that I wished there were more remains that I could give him, and instantly realizing what a strange thought that was.

I wondered just how much pain one man could stand. First he endured the fear and uncertainty of having a missing daughter. Then, on Father's Day, of all days, he hears that the body we had found could be that of his daughter. Then it is confirmed that she had been murdered and her bones scattered and eaten by a dog. He obviously is a man of strong faith.

He was quoted in an interview as saying, "If Christine is alive, she will walk toward me. If she's coming from the other path, someone else will bring her to me." It was with great sorrow that I brought her back to him, in a small body bag.

To this day, we do not know who murdered Christine Moore. We still talk of her case, and our guts tend to knot up a little when we cruise by the church now and then. But her murder remains unsolved. Of course, that's the official stance. I, for one, would sure like to hear the truth from Derrick Lee about this case, but he ain't talking.

Whaddya expect, Lou. He's a sociopath.

ARE WE SAFE?

Neither Geralyn's nor Christine's death was linked to Gina Green's. Then came the call. It was May 31, 2002. "We have the body of a white female student on Sharlo Avenue. It's over near LSU."

I know where it is because my older sons lived nearby when they were in college. I am also informed that a roommate discovered the victim's body at about two P.M.

It is a beautiful summer day. The oaks in the area are offering ample shade to the crowd that had gathered at the little grouping of student townhouses near the LSU campus. Inside the first townhouse to the left is the body of Charlotte Murray Pace. A person poised to make a difference in our world, Pace was the youngest person in campus history to earn a master's degree in business administration. She was a vivacious young woman who enjoyed life and was on her way to making a contribution to our society. She was well liked and family oriented. But all that was moot now. Her potential had died with her in a violent struggle that ended in rape and mutilation.

The lawn between the townhouses is crowded with various police officers, local PD crime-scene techs, state police crime-scene investigators, and our ubiquitous coroner van. I had been forced to relinquish my tired-out old Green Hornet to the government auction lot and was now driving a newer black Crown Victoria, which I parked under one of the oaks near the road.

As soon as I exit the car, I hear my name being called out. The voice is that of a longtime friend who lives in the area.

"Lou! Lou, over here!" Kandi runs toward me. "What happened? I hear a girl was killed. Are we safe? What do we do?" She's scared. It's an appropriate response, given the situation.

I tell her that I will get back to her.

De is with me as I walk up to the townhouse. The detective tells me we have a bad situation—one of the worst he has ever seen. I wait outside. There is no rush. I want them to have everything processed—except the body—before I go in.

There is a white wooden swing on the little porch. I enter and notice a small plate on the sofa arm, the one farthest from the door. There is a half-eaten sandwich on the plate and half a bottle of Diet Dr Pepper on the table. She was eating when he first confronted her. *Did she open the door, or was the door unlocked?*

There is a small amount of blood on the carpet. The attack started here, by the couch. It makes me think the door was not locked and he barged in, but that is supposition. In any case, it was more than he counted on—she fought back. The blitz had not worked. *Did she deflect the blow? Did he miss?*

I follow the fight trail into a small hallway to the right. The drywall is caved in at the lower third of the far wall. There is evidence of smeared blood. *Were you trying to escape? Were you going for a weapon or just trying to get away? Did he catch you and bang your head against the wall? That didn't stop you, did it? You fought your way into the bedroom. Brave girl!*

The whole bedroom looks crimson, reflecting the ugly

bloodbath that took place here. She is on the floor, near the door, on her back. Her head is nearest to the door and her legs are apart. I cannot see her eyes due to the swelling. She is nude from the waist down with her bra pushed up over her breasts. Her throat is cut deeply—there are several slashes to the area. *Overkill. Rage.*

As I step closer, I can see that her body is riddled with dozens of stab wounds, to her face, eyes, ears, and other parts of her body. Some are blunt at both ends—*screwdriver! Did you try to defend yourself with the screwdriver and he wrestled it from you? What about the knife? You have defense wounds on your arms. You tried to block the blows from the knife and the screwdriver. Did he take a break once you were unconscious and go to find the knife to cut your throat? What are you telling us?*

In conjunction with the state police crime lab techs, we collect the evidence with painstaking precision. We have no concept of time. We are frozen in the moments of her death. We are there with her as best we can be. We rely on time-tested techniques, and with our alternate light source we gather the trace evidence from her brutalized remains. It will seal the killer's fate. *I cannot shake the image of the crimson room.*

De tells me she feels proud that Murray put up such a fight. *She identifies with the victim—that could have some adverse effects on her later, but that will have to wait.* Surely she marked him. *Yes. He must have been a bloody mess when he was through. How did he just walk out like that? Even more amazing, considering the horrific life-and-death struggle that went on here, is that the neighbor, who was at home at the time, and whose townhouse shared a common wall with Murray's, did not hear anything.*

De is standing against the wall nearest the door when she makes that prophetic statement: *"He's done it again. First it was Gina, now her."* Those words still ring in my ears. De, surrounded by seasoned investigators, finds that her opinion is pretty much discounted. She is even told, patronizingly, why her supposition is flawed. If there were ever two diametrically opposed scenes, these two are it. This crime scene is bloodier and different from anything that I'd seen to date—so much so that admittedly, I did not connect the two murders at that time.

The sort of behavior on display here seemed so unlike what we'd seen in Gina's murder. From what I'd read in the serial killer books, I expected the same modus operandi. I'm reminded of a lesson I had along those lines in medical school. While discussing a surgical case with one of my professors, I commented that the patient's symptoms did not follow the book. My professor's response was: *"I guess he didn't read the book."*

But DeAnn made the connection. I'm not sure how she knew—female intuition, I guess. Had we known the details of Geralyn DeSoto's murder and the crime scene, we might have been more open to De's input at the time. But we were not, and, unfortunately, there is not much communication across jurisdictions.

We place Murray in a fresh homicide sheet that will retain any trace evidence. It's almost like putting her in a burial shroud of old.

There were bloody footprints in the house, so the crime-scene tech was out front photographing the soles of everyone's boots who had been in the house. Was it crime-scene contamination from one of us tromping through the house or did the killer leave his tracks? We emerged into the sunlight and the staring eyes.

We escort Murray to the van and on to the morgue. There we will listen even more intently to what she has to tell us. I walk over to inform the detectives about the time of the autopsy. Somewhere in the background I am aware of some residents in the area telling the police about a suspicious dark-skinned male who was hanging around and looking at Murray's townhouse. It's police business. I head off for the morgue.

During the autopsy, we would find that Charlotte Murray Pace had sustained eighty-one stab wounds. Many of them were blunt on each end, indicating that a screwdriver was indeed one of the weapons used on her. Murray had just moved into the townhouse and there were tools about. A blue, foot-long, flat-headed screwdriver was found at the scene. She was also stabbed with a knife. I suspect he may have gotten it, too, from her house. He stabbed her repeatedly about the face, neck, chest, and hands.

On July 9, authorities announced that it was indeed the same man. DNA evidence obtained from the murder of Charlotte Murray Pace proved to match that obtained from the Gina Green case. A monster was loose in the city. DeAnn was right. *"He's done it again!"*

This tragic news confirmed our worst fears: we had another series of female murders. A different series, the major difference being the victims' profiles. These were not black prostitute homicides, these were attractive, decent, career-oriented women;

women who did not put themselves in harm's way. Other women in the community identified with them. Not only was there public outcry but there was also public outrage and panic. This vile creature who killed and raped these women would become known as the Baton Rouge Serial Killer.

As more information came in, the links between Gina and Murray grew stronger, and maybe that would lead us to who killed them. Of course, the greatest clue, the dark-skinned man seen at Sharlo, had evidently fallen by the wayside, or under the Sharlo oaks. It was certainly significant to learn that Murray had moved to Sharlo Avenue only two days before her death. Oddly, her previous address was only three doors down from Gina Green's home on Stanford! Also missing from Murray's house was her cell phone.

When I was a kid, my grandma used to say: "The acorn doesn't fall far from the tree." Perhaps the wisdom of that saying did not really hit me until I visited with Murray's mother, Ann Pace. She's an attractive lady with style and poise who had her world shattered. But she's the kind of woman who picks up the pieces and does what she has to do. I remember it like it was yesterday: We were in my office, Ann, her daughter Sam, and me. Ann wanted to know everything about her baby. Grief is a complex issue and different people handle it in different ways. Some folks don't want to know any specifics and choose to remember their loved ones as they were, not as they died. Others want to know every detail so they can experience the pain their loved one went through. It is a way of being there for them. Ann chose the latter way—she wanted to know. So I went over the autopsy with Ann Pace. It was tough for both of us, but much tougher for her. Still, it hurt to see her in so much pain. The clinician in me realizes that you cannot take away people's pain; you must let them have it. But I also realize that honesty without compassion is brutality. You can only be there for them during the process. Her other daughter, Sam, has been there for her and is "cut from the same cloth," as my grandma would say. As I have come to know more about Murray, primarily through her mom, I'm sure she did put up a valiant effort to thwart her killer, and that she never surrendered or gave up. He killed her body but not her spirit. That must have enraged him even more. Regretfully, I came to know Charlotte Murray Pace only posthumously, through her family and friends, but I see where she got her courage.

I think Ann Pace summed it up for all of us in late April of 2003: "There is nothing left to seek but justice."

TAKEN

On July 12, 2002, three days after the DNA link between Gina and Murray was announced, Pam Kinamore, dedicated mother and wife, was abducted from her Briarwood Place home. She lived with her husband, Byron, and their twelve-year-old son in an upscale neighborhood off of the southern part of Airline Highway. Pam was an attractive dark-haired lady who grew up in New Orleans and put herself through college with the aid of money she won in the Miss Jefferson Parish beauty contest. A decorator, Pam owned her own business, an antiques store called Comforts and Joys, in the small town of Denham

Springs, Louisiana, which is just over the Amite River from Baton Rouge. I say all this because the crime showed no apparent LSU connection at the time. Gina and Murray lived near LSU; Murray was a student there; lots of career-oriented women walk around there on a daily basis; thousands of them, in fact. What better place?

When Pam's husband came home from work later that evening, her car was there, but she was not. He waited for her to show up and became worried as time passed. It was not like her to disappear without saying where she was going. Later that evening he called the police and reported her missing. Her mother, Lynne Marino, drove in from her home in New Orleans and was appropriately distraught over the situation. There is still concern among family members that the police tended to discount Pam's disappearance during the first critical hours of her abduction. The sheriff's deputy even opined that she might be missing because she was having an affair and that there could be a simple explanation for her absence. The family knew better.

Four days later, a state survey crew working near Whiskey Bay Bridge discovered a gruesome sight. Whiskey Bay, so named because an old riverboat full of whiskey sank in the area, is located on the I-10 corridor that runs through Baton Rouge and on to Lafayette and farther westward. And there, located in a marshy area of the woods beneath the bridge, surveyors came upon a human-like figure lying near the water's edge. As they approached, they discovered it was the nude, decomposing corpse of a woman. It was Pam Kinamore. Her throat had been cut.

Since her body was discovered in Iberville Parish, her murder fell under the jurisdiction of that coroner and that sheriff, and it was they who processed the crime scene. They also collected a piece of evidence that seemingly had no overt connection to the crime. It was a piece of telephone cord.

The Iberville Parish coroner sent Pam's remains to New Orleans for autopsy. Later, when the East Baton Rouge Parish district attorney, Doug Moreau, assumed jurisdiction, she became my responsibility and we brought her home. I had a forensic anthropologist as well as our own forensic pathologist waiting for Pam, to reexamine her remains.

Pam's body was transported from New Orleans directly to the LSU forensics lab. My forensic anthropologist, Mary Manheim, again was tasked with the examination of the bones. Much to my shock, Pam had not been X-rayed at the New Orleans morgue. Subsequently, we subjected her remains to total body X-ray to see if there were any clues, such as bone fractures that might indicate a certain type of trauma, or nicks on her ribs that might indicate stab wounds. This was necessary, as her remains were badly decomposed when I got her. I was also looking for a skull fracture that might indicate an attack to her head. Once she was X-rayed at the FACES lab at LSU, I returned her to the trailer morgue for a second autopsy/review by Dr. Michael Cramer, my forensic pathologist who does the actual autopsy.

It's not that I had any reservations about what the New Orleans pathologist had done on the gross autopsy. He's very well thought of professionally. I just wanted to make sure everything was done to my standards and protocols. I had already examined two of this killer's known victims, as well as Christine Moore, and there might be some subtleties that we could see or know to look for that someone else might not. The jurisdictional issue was also addressed by establishing a working relationship between my office and that of the Iberville coroner, just in case a similar situation arose again.

Pam's wedding ring was still on her finger. We carefully retrieved it. Obviously the killer was not interested in robbery. His heinous appetites were far more depraved. Pam's ultimate cause of death was exsanguination—she bled to death.

On July 29, two weeks after Pam's body was discovered, the Louisiana State Crime Lab was able to link her death to the killer of Gina Wilson Green and Charlotte Murray Pace.

As she was the third "linked" victim, it was announced that her killer had fulfilled the requirement for being officially labeled a "serial killer." Terror seized the city.

Pam's death hit especially close to home for my family and me. Her son goes to Parkview Baptist School, the same school as my eleven-year-old son. Pam's son is a grade ahead, and DeAnn knew who Pam was—another school mom.

It fell to me to bring Pam's wedding ring back to her husband. What is there to say? Not much except "I'm sorry" and "we're trying to find the guy." Little solace to a man who just lost his loving wife and the mother of his young son. Even worse, the cops had considered him a suspect in the first days of the investigation. *How does a man in a wheelchair carry a body through a muddy swamp? He'd been injured many years ago and was wheelchair-bound even before they were married. Gimme a break!*

As I was walking back to my car from the front door of Byron Kinamore's house, which sat on a street lined with magnolias, I was accosted by a news reporter. I guess they had the house staked out. I was enraged that the media would encroach upon this family in their time of acute grief. I showed my displeasure in my choice words and the reporter got the message. Needless to say, there was no interview. Someone with me at the time told me he'd never seen me that angry and thought I was going to hit the reporter. Maybe I was too close to the situation. And maybe not. Maybe it was the reporter who was too close.

I was not at the initial Whiskey Bay scene due to jurisdictional issues, but I got to know Pam's family and we did visit the area with them. A weathered wreath, placed there by her family, marked the spot where Pam was found. I credit her mom, Lynne Marino, with keeping these deaths in the forefront of everyone's minds and keeping the focus on the investigation. She is a most tenacious and outspoken lady, and she demanded accountability from the authorities. She took a lot of flack for that, but she never let up. And people listened. And, more important, they responded. The families became the conscience of the investigation. They insisted upon actions and answers from the authorities, they got politicians involved, they held rallies, and they also met and did their own victim-profile analyses. In this way did it emerge that Pam had been at LSU for a function the day before her abduction. Maybe there was a connection. Maybe this was a clue as to the predator's hunting ground.

ELEVEN

Unsolved Mysteries

Random murder makes people feel very vulnerable, especially when they share multiple traits with the victims. It scares you for your wife, your children, everyone around you. For instance, what if Pam Kinamore's young son had been in the house? What would have happened to him? Fear is a protective emotion. Fear is not a bad thing, but ignoring it is. DeAnn was listening to her fear in early August when she had gone to visit a friend at Our Lady of the Lake Hospital. She called me at nine P.M. to come escort her to her car because there was no one around to do it. I drove over to the hospital and walked her to her car. I guess I could have called hospital security, but I preferred to handle this myself. What mattered was that she was fearful of venturing out into a dark lot because of the serial killer.

Then there was the evening when she called to tell me that Jack, our three-and-a-half-pound Yorkie, was barking and there might be someone lurking about. DeAnn is no wimp and she knows how to handle a firearm. Yet she had grabbed my old antique .45 Webly revolver instead of her own .357 Smith & Wesson and was grasping it in her hand while standing in the hallway waiting for the intruder—and she was scared. I rushed home and cautiously announced my presence. I didn't want to get shot. Actually, I am absolutely certain that if there had been an intruder, it would have been his dead ass that I'd have been greeted by. I understand her feelings; as coroner I am in a highly visible position. She had been to those death scenes and understood just how ruthless this bastard was. What's more, she had been on camera and knew he was watching the TV. Also—she fit the victim profile.

At any rate, all of Baton Rouge was ready for the police to do something. And, we all readily embraced the belief that a task force was that something. In August 2002, a task force was formed of some forty investigators from various law-enforcement agencies, including the FBI, the Baton Rouge Police Department, the East Baton Rouge Parish Sheriff's Office, the Louisiana State Police, the Lafayette Parish Sheriff's Office, and the Iberville Parish Sheriff's Office, among others. This Multi-Agency Homicide Task Force was formed to establish clearer lines of communication among the various bureaus. They believed this could help them share information better and catch the killer more quickly.

The serial-killer cases were top priority now, and no expense was to be spared. Lab results would be processed in a timely manner. That was good news. Of course, it did little for all the other cases they had to process. The dedicated scientists at the Louisiana State Crime Lab continued to compare DNA samples of the killer with samples taken from several dozen unsolved murder cases over the previous decade. Limited financial and personnel resources made this an arduous task, to say the least. DNA casework was already backlogged. One of the reasons for that backlog was me—and other coroners. We do rape kits on living and deceased persons and we submit them to the crime lab. We don't do our own work for the simple reason that the crime lab does it at no cost to us. When I tried to get funding to do some of my own lab in house, I was informed that we got it for free, so that was not a sound move, in fiscal terms. The end result is that just about everyone in the area sends samples to the state police crime lab, fully aware that it might take months to get the results back, depending on the priority of the case.

The response to the linked deaths of Gina, Murray, and Pam was dramatic. FBI experts came to town and developed the killer's profile. A twenty-four-hour hot line was set up. Hundreds of males were swabbed for DNA. The families of the victims became outspoken advocates for the killer's capture. A bounty was placed on his head. National news media came to town. Special bloodhounds were brought in. There were waiting lists to purchase Mace spray. Self-defense seminars flourished. Gun sales increased dramatically.

Sadly, the public's expectations of the FBI profilers was a little overblown. This was colored in part by books and movies like *The Silence of the Lambs.* I think people expected the FBI to be able to give us the guy's name, address, and Social Security number, and the name of his maternal grandmother to boot. Such high and unrealistic expectations introduced an element of dissatisfaction into the reality of the process.

The tip line was a great idea. Historically, many serial killers have been caught thanks to information supplied by people in the community. However, the robust community response quickly deluged the system. In short, it was swamped. There were lots of tips about white pickup trucks. A witness came forward saying he had seen a nude woman slumped over in the passenger seat of a white pickup truck pulling off the Whiskey Bay exit of I-10. Under hypnosis, he said the truck was white and so was the driver. Suddenly, if you were white—and drove a white pickup—there were more than 27,000 in the Baton Rouge area, according to one account—you were eyed with caution. Prioritizing those tips became an issue. The truck in question was described as a late-1980s GM truck with bad paint and bad window tinting. It wasn't too long before bumper stickers began to appear on some white pickup trucks that read simply: NOT THE SERIAL KILLER.

Even DNA became a "sore spot." If a person was turned in to the tip line, that person got swabbed. Sounds innocuous enough on the surface, doesn't it? Unfortunately, it seemed to take on a life of its own and became a substitute for police investigations. Several "suspects" felt they were being coerced by Gestapo tactics. Here is the scenario. Two police detectives show up at your door. They tell you an anonymous tipster thinks you might be the serial killer and they are here for a DNA swab. They don't check to see if you were in town or even in the country at the time of the murders. They just want the DNA. If you resist, it is a sign of guilt and they will get a subpoena, which will be noted in the public record, letting everyone know you are a serial-killer suspect. If you comply, you don't get the negative result back when you are ruled out, and you wonder later what database your DNA "fingerprint" will end up in. Not surprisingly, lawsuits followed.

I felt obliged to find out exactly what was going on at the safety seminars that were being held throughout the city, so I attended one. *You can't support something unless you know what it is you're supporting.* Since I was there to observe, I found myself standing in the back of a library that was jam-packed with women of all ages. The safety tips were good, and the presentations by peace officers were on target. The one thing that bothered me a little was the self-defense demonstration on how to handle an attacker. The instructors have much more confidence in the simple tactics they were teaching than I did. I figured the killer to go two hundred pounds or so and be strong as an ox. There was no way these young women were going to fend him off in a physical altercation. Charlotte Murray Pace had given him a hell of a battle but died anyway. So I watched the demonstrations and hoped the women were more attuned to safety and were not getting a dose of false confidence.

At least most women no longer jogged alone.

One of the most "colorful" comments relating to self-defense came from our governor, Mike Foster. During one of his *Live Mike* weekly radio shows, he told the citizens of Louisiana that "if you have some fruitcake running around . . . [a gun] sure can save you a lot of grief." That view seemed to be shared by lots of folks, because in the first week of August 2002, the state police had 447 requests for concealed handgun permits. That averages out to about ten per hour. Now, I'm a believer in appropriate handgun carry. I'm comfortable around firearms, and I know my way around a gun shop. It was at a local gun shop that I became somewhat alarmed by what I saw and overheard. A dealer was trying to convince this 110-pound young woman that she needed a .357 Sigarms pistol. Now, there is no way this novice was going to be able to control that weapon, at least not without considerable training. And I'm sure she'd rapidly get tired of carrying such a huge gun around. Maybe the salesman was just uninformed, or maybe he was trying to sell an expensive gun because his small handgun stock was low. I hope it was for the former reason.

Rumors of all sorts were rampant and this fueled the sense of terror. It's important not to get caught up in that emotional roller coaster. If you become so emotionally involved with a murder victim that you let your emotions impair your objectivity, you can make mistakes that could allow the perpetrator to get away or get a less appropriate sentence. For instance, if I'm overheard at a crime scene saying something really stupid like: "I'd like to kill the dirty, gutless, sociopathic son of a bitch who did this," I have compromised myself for later testimony. "Is it not true doctor that you said in an open forum that you would like to kill my client? . . . Do you know my client? . . . Can you really be objective here? . . ." And, then, if you jump to conclusions, you might end up trying to make the facts fit your pet theory, ignoring other evidence that doesn't fit your picture of what happened. Of course you'll feel emotions at the scene, but you must compartmentalize them and put them aside and deal with them later.

And in the midst of it all, the "usual" flow of murders, accidental deaths, questionable deaths, and suicides continued. There were the business-as-usual homicides also.

TWELVE

In the Sights
of a Sniper

At about 6:35 P.M. on September 23, 2002, Hong Im Ballenger locked the front door of the Beauty Depot, her beauty-supply shop located near the center of Baton Rouge, and began walking to her car. In one hand she held her keys; in the other, her empty lunchbox. A stiff breeze was blowing, a welcome relief from the unseasonably warm weather.

A witness told police that as the forty-five-year-old Korean woman was about to get into her Mitsubishi Montero, a man confronted her. Thus did the police get the initial impression that someone came up to her, possibly demanded her purse, then shot her and ran off. Another witness would later report that a dark-blue car parked in a vacant field across from the parking lot approached the spot where Mrs. Ballenger was shot and picked up the man holding her purse.

She was shot once in the head, killing her almost instantly—not even enough time to pull her can of pepper spray from her pants pocket. She obviously hadn't been expecting any trouble, as the spray canister was deep within her pocket. That might suggest that no suspicious person had been in the store earlier, or at least no one who alarmed her. The murder occurred against the backdrop of a serial killer on the loose in Baton Rouge. Stores were selling out of Mace, and self-defense classes were filled. Public-service TV spots ran nonstop. None of that does any good if you don't stay on full alert, but, as we would learn in time, no amount of caution could have saved Mrs. Ballenger.

When I walked up to the crime scene, the wind was blowing hard enough to lift the plastic blanket off of Mrs. Ballenger's body. Her shop was on the north side of Florida Boulevard, about two blocks from the Baton Rouge General Hospital. I know the area well because I worked at the General for about twenty years. The shop sits on the cusp of a rough neighborhood that has hopes of being improved by mid-city revivalist efforts.

Across the street, on the south side of Florida Boulevard, the inevitable crowd was building and people began climbing on the hoods of their parked cars to observe the show. They were joined by the local news crews, and the cameras were rolling. The dome strobes from the official vehicles only added to the scene's slightly carnival effect.

There was a long, partially coagulated trickle of blood running out from under the blue tarp. It had followed the slight downhill grade of the blacktop parking lot and pooled up about three or four feet from the edge of the improvised body cover. The source of the blood was Mrs. Ballenger's head and neck area. Her body was stretched out parallel to the car, her lunch container at her side. I examined her visually only, careful not to touch her lest I contaminate any evidence; plus, the homicide detectives needed to see how her body was before I moved her.

As I stood up from my initial, brief visual examination of Mrs. Ballenger, I noticed a child in a nearby truck—I soon found out it was her son. Night had just fallen and the boy's face was illuminated by the staccato strobe of the police vehicle lights. He looked so small and vulnerable. He never looked my way. It struck me how confused and scared he must be.

Mrs. Ballenger's husband, fifty-five-year old Jim Ballenger, was appropriately distraught. *It's a shame we have to assess that, but we suspect everyone, especially spouses, in any homicide.* His wife was a native of Inchon, South Korea, and they had married when he was in the military. They moved to Baton Rouge in 1996 with their three sons. "I know my wife is in heaven," he told the local press later. And he did not believe in the death penalty. "Jesus said to forgive, and I am born again. The man who did it needs to do time in jail."

One of the things that hit me the hardest at this particular crime scene was the murder victim's son. He looked to be about the same age as my youngest son. He seemed so small sitting there in the huge, light-brown Ford pickup, which was one of those dual cab types with a long wheelbase. He was sitting there, staring straight ahead through the windshield, eyes focused on nothing in particular. I guess he was just trying to avert his gaze from the horror in the parking lot to his right. The truck was barred from entering the lot by a ribbon of yellow crime-scene tape that sputtered as it resisted the incoming cool front.

I surmised that he knew this parking lot well. His mother worked here at the Beauty Depot, and this lot served her customers. If he's anything like my youngest, Michael, he'd found occasion to roam about the lot, perhaps when he came to meet his mom after work on their way to eat out. Maybe he had even ridden his bike or skateboard up the ramp there, or over those bumps. And now his mom was there under the tarp and she wasn't coming home. *Maybe I project too much.* The wind kept trying to blow the blue blanket up, but a man in a dark shirt quickly put his foot on the edge of the blanket to keep it down—shielding what was beneath. The child's only hope was to look straight ahead.

There is no way to make any sense out of such malicious violence. I feel utterly powerless. I was shaken back to the reason I was here by one of the uniforms. He tapped me on the shoulder as I had not responded to his initial request. "Hey, Doc, the PIO wants you to talk to the husband of the deceased." The uniform spoke quietly and motioned to where the public information officer anxiously awaited my presence.

The streets can be very unforgiving. My assessment of this situation started as soon as I began to walk over to Mr. Ballenger and the public information officer, Corporal Don Kelly. They were only about fifteen yards away but the distance seemed like miles. A deluge of questions flooded my brain:

What is there to say? How much is he ready to hear? How much does he want to hear? How much can I tell him?

I know words cannot even approach an understanding of the magnitude and depth of their loss. Death is a very personal thing, and so is grief. Maybe as a doctor I want to take away the pain. Seems grandiose, doesn't it? Maybe being part of the judicial branch, I feel some guilt that this happened. Maybe it makes me feel vulnerable myself, and for my family, and I know words cannot ease a loss. So why say them? Because that's all I have to offer at the time!

I've seen the full gamut of responses from family members. Some are just numb, or in shock. They simply cannot process the horror. I've also been physically attacked during a notification—a case of what we call displaced rage. When I was an ER doc back at the General in 1977, an elderly black female died of congestive heart failure. I had the whole family (*wrong*) assembled in a small office (*wrong*). I went in and closed the door behind me (*wrong*). Then I tried to break the news. There must have been twenty people packed into that room—we're talking *very* close quarters—all sitting and standing, like they were in bleachers and I was the main attraction. So I have my back to the closed door that opens inward (*wrong*) and I say something like "She did not make it . . ." (no response) ". . . we tried our best but her heart just gave out . . ." (no response) . . .

Then a matriarch sitting directly in front of me asked, "You mean she's *dead*?" I said, "Yes, she is dead." The matriarch shrieked and threw an umbrella at me—point first. I dodged the projectile and it bounced off the door next to my head. I noticed as I made my escape that it left a dent in the door. It was a lesson I never forgot and one that was always in the forefront of my mind at times like this.

The PIO introduced me to Mr. Ballenger. I shook his hand. His eyes told me all I needed to know. He was in profound distress but he was doing what he needed to do. We respect that out here.

I remember putting my hand on his shoulder. It's about empathy, I guess.

When his eyes met mine I saw that questioning pain I have seen too often. Any fears I may have had about a violent response on his part were dismissed.

"I am so sorry for your loss." *The words are so inadequate.* "How can I help?"

He wanted to know how she had died.

That old pearl of wisdom guided me again—*Truth without compassion is brutality.* It's more than a pearl, it is an extrapolation right out of the Hippocratic Oath ("First, do no harm . . .").

I have learned that it is usually better to say things straight out and be supportive. It was apparent that he wanted and needed to hear the facts. So I told him.

"Your wife sustained a single gunshot wound to her head, and she died from that injury. There is no other sign of trauma or assault." *There is no easy way to say these things.*

He nodded in the affirmative. When a person really loves another, there is always one question that they ask right away. "Did she suffer?"

"No sir, she died instantaneously. I doubt she even knew what happened."

There was no need to go into how I deduced that information.

Truth without compassion is brutality.

"Thank God for that!" he responded, and I knew he meant it.

He turned to his son, who was still staring straight ahead. People came to help—the family minister and some other folks—and after a brief exchange, the child, who remained virtually motionless, was taken away. *I felt some relief that he was gone from this nightmare, yet I knew this night would never end for him.* Mr. Ballenger told me that family friends and the minister would look after his children.

There are times when people just need permission or guidance from someone in authority to be able to do the right thing for themselves and their family members. Nobody is ever prepared for something like this. They don't really know what to do or what is going to happen. I informed him that we would be doing an autopsy and that the funeral home of his choice should contact us concerning the release of his wife's body.

Even as I was saying this, I was not sure how much he was absorbing. The only thing that I really knew was that the look on that child's face would intrude on my own daily thoughts for many years.

The homicide detectives were late in arriving and did not make the scene until an hour or so after the call-in. That's a long time to wait around and leave someone's body under a tarp. It's an uncomfortable situation, with the potential to magnify the trauma for the family. It also gives the crowd more time to gather, and that can mean trouble of another sort. Old enemies come down from the northern part of the neighborhood and run into one another. And they may have scores to settle. We don't need stray bullets and another crime scene. The detectives were probably involved in the search for the serial killer—especially since this was the one-year anniversary of the first known victim's death. But people expect a rapid response, and they were not getting it.

I hate having to leave the victim's body out in a parking lot, but I absolutely understand that the detectives need to see the undisturbed scene. So we waited and it seemed like an eternity. While we waited, Adam and I discussed options and possibilities. Adam was a young, enthusiastic crime-scene technical officer and had just returned from an investigator school on crime-scene reconstruction. He's a damn good investigator and we had grown to respect each other and worked together well on death-scene investigations. We knew that the FBI still had its bloodhounds in town for the serial-killer search and wondered if the dogs could be of service here.

The detectives finally arrived and I was able to get back to the awful task at hand. The bullet appeared to have entered the victim's left lateral upper neck area and exited through the mouth. Her lower jaw was pulverized, and teeth were scattered about the pavement. The bullet then slammed into the driver's-side mirror of the SUV and lodged there.

She was so small—maybe 110 pounds—and there were several pints of blood on the ground. I surmised that she was hit at the upper cervical spine, probably suffered instant paralysis, and bled to death. Her purse was missing and there were no empty shell casings about—shot with a powerful revolver, was my guess. (I was wrong.) Her keys were still in her right hand and there was no sign of any effort to break her fall by putting her hands up.

These are challenging times for me. The need to be professional, while the reality of the damage that has been done to this family and child as well as his siblings make it difficult. But the job has to be done right if we're going to get the guy who did this heinous thing.

I examined the spray of Hong's blood and tissue that had been blown onto the vehicle. The fine mist of particles indicated a high-velocity projectile—I was still banking on a .357 as the murder weapon.

As I examined Mrs. Ballenger, I inquired if she had been moved at all. No one had an answer to my question.

It's an important question. In the event emergency first responders moved Mrs. Ballenger's body, we needed to know the position she was in prior to that movement. If the blood splatter doesn't fit the way her body is positioned, somebody moved her. If the perpetrator moved her, then we have a greater chance of finding trace evidence from the killer—trace evidence he left behind that will link him to the scene and help convict him.

So I decided to go to the source. "Adam, do me a favor and radio EMS to see if they moved her." He was already on it. They had not moved her to any extent other than to verify she was dead. That was good news. I depend on EMS to stand for Emergency Medical Service, not Evidence Meddling Service. I must admit, though that the EMS guys in East Baton Rouge Parish are pretty good about not disturbing evidence—most of the time, anyway.

A visual replay of how the injury must have occurred played through the right side of my brain while the analytical functions of my left brain processed the information. *The bullet smashes into her neck, shattering vertebrae; then fragments of bone and the bullet itself sever her spinal cord and various blood vessels in the area. Then it exits through her face. She never had a chance. She was minding her own business, perhaps thinking about the best route to get home to her family—then a split second of shock and it's over. Over forever. No time to prepare, no time to say goodbye, no time to defend herself from the cowardly bastard who did this.*

As I have mentioned, this woman's murder occurred during the hunt for the Baton Rouge serial killer. As part of the efforts of that task force, the FBI had brought in three scent-tracking bloodhounds to the city to search for serial-killer evidence. I did not understand all the intricacies of bringing them in and they seemed to be of little use as far as getting the serial-killer mystery solved. *But no expense was spared.* However, this seemed like a lucky coincidence, and the BRPD requested that one of the dogs be dispatched to the Beauty Depot to pick up the trail of Mrs. Ballenger's killer.

The dog was offered a bullet fragment that had gone through Mrs. Ballenger's neck and face, then given a chance to smell her body through the body bag. The dog then smelled everyone in the area and charged off in the direction in which the killer had reportedly fled, a group of officers and dog handlers trotting off after the animal. I had my doubts that the bullet fragment would offer a scent. The idea that a bullet that passed through a human and lodged in a car could have the scent of the person who put the cartridge into the gun seems like a helluva reach. But I'm no dog expert.

Incidentally, the dog did not find the murderer. I'm not sure who, if anybody, they found at the end of the scent trail but I'll bet he was surprised.

A month later, first-degree murder warrants were issued for John Allen Muhammad, age forty-one, and Lee Boyd Malvo (alias John Lee Malvo), his seventeen-year-old companion, infamous now for their killing spree that spanned five states.

Mrs. Ballenger happened to be one of their first victims. On the lam, they faced multiple state and federal counts in Alabama and Louisiana, as well as charges in the sniper spree that left ten people dead and three others wounded in Maryland, Virginia, and Washington, D.C.

Muhammad and Malvo were finally arrested in connection with the Beltway sniper attacks early on the morning of October 24, 2002, at a Maryland rest stop where they were sleeping in their faded-blue 1990 Chevrolet Caprice.

Hidden in the Caprice was the Bushmaster .223-caliber rifle, which would be positively linked to the Ballenger killing by ballistics comparisons. It was the same weapon used in eight of the ten D.C.-area sniper killings.

Receipts found in the suspects' Caprice also put them in Baton Rouge on September 23—the day Hong Im Ballenger was killed. Muhammad was a local boy; he grew up in Baton Rouge. Family members said that he had recently passed through town with Malvo. While they were here, they reportedly went to the YMCA to exercise—the same Y where my wife and I work out. Who knows, they could have been in there at the same time we were. Think about that for a moment. You never know how close you may be to such incarnations of evil. Oftentimes they simply blend in.

Justice was finally served. The ultimate plan of the two killers was to extort $10 million from the government in exchange for an end to the shootings. Muhammad's trial began in October 2003, and the following month he was found guilty. Four months later he was sentenced to Virginia's death row, where he sits now, awaiting execution. A jury convicted Malvo of capital murder on December 18, 2003; he received a sentence of life imprisonment. He received a second life term in October of 2004.

Hong's husband is more forgiving than I could ever be. He obviously has a strong anchor in his faith. Mr. Ballenger has been quoted as saying he wanted the killers to get life in prison—that way they will have a chance to repent and go to heaven.

Hong's sister is not so forgiving. As far as she is concerned, sparing the lives of such malicious men only gives them the chance to kill somebody else while in jail. *That somebody could be a guard.*

I'm of the same ilk. I still see the look on that child's face in my dreams. I see me talking to his dad, and the young boy staring straight ahead, not wanting to know what I was saying and not wanting to look into the parking lot. I wonder how that child is doing tonight without his mom. I wonder about the afterlife. I wonder if Hong is looking over him right now. I wonder about what all this means, and I wish for the death of the perpetrators.

I have anger, pain, and outrage. Our only solace is in the fact that the two despicable perpetrators have been caught and will ultimately face justice. The district attorney here in East Baton Rouge Parish is waiting for our turn to try Malvo and Muhammad. If you murder someone here, you don't just walk or run away, as the case may be. That dog just don't hunt.

THIRTEEN

A Killer
Strikes Again

TRINEISHA DENÉ COLOMB

And so in the midst of the Baton Rouge Serial Killer ordeal, Malvo and Muhammad were captured and those two serial killers were off the streets. But the man who was credited with the murders of Gina, Murray, and Pam was still active. Evidently the Baton Rouge heat was getting to him, as he struck next near Lafayette, Louisiana, a town about fifty miles west of Baton Rouge, along Interstate 10. The drive time between the cities is an hour or less. This time he killed an attractive and successful twenty-three-year-old black female, Trineisha Dené Colomb (known to family and friends as Dené). She had been in the army for two years, loved her country, and had plans to become a Marine.

Dené became a missing person when she disappeared on November 21, 2002. A local resident noticed her car parked on Robbie Road that same day but did not think much of it at the time. When it was still there the next day he reported it to the police. Her black 1994 Mazda MX3 had the keys in the ignition. Her coin purse and license were also found inside the vehicle. A search of the area ensued but was not fruitful. She was still a missing person. Her family guessed that she must have been in the area to visit the grave of her mother, Verna, who had died of cancer seven months earlier. Dené visited her mother's gravesite frequently.

Two days later, she was no longer a missing person, she was a murder victim. A rabbit hunter discovered the nude remains of Ms. Colomb several hundred yards off the roadside in a wooded area in St. Landry Parish twenty miles from where her car had been discovered. An autopsy revealed that she had been bludgeoned to death and died from blunt trauma to the head. I've met some of her family members at memorial rallies, and it was most apparent that they were not going to let this just go away. But due to jurisdictional boundaries, I was not officially involved in her case.

Two weeks after Dené's body was discovered, a caller came forward with news that a white pickup truck had been seen parked behind her car on the day she disappeared. The occupant was a white male thirty to forty years old and had an "intense and intimidating stare." A sketch of the man from the white truck was widely circulated by the task force. He was a "person of interest" and not a suspect per se. A new amended profile was developed and released to the public. This "person of interest" reinforced the "white male in a white truck" image of the killer.

One month after her death, DNA evidence would indicate that Dené was killed by the same man who killed Gina Wilson Green, Charlotte Murray Pace, and Pam Kinamore. On December 24, 2002, the *Baton Rouge Advocate* ran the headline: BR KILLER STRIKES IN LAFAYETTE. The mythical "racial barrier" that serial killers are not supposed to cross had been broken.

This abduction was also a deviation of the killer's mode of operation. To date, he was tagged as liking home abductions. What was he doing in a graveyard? Yet, there had been another attack in another cemetery some years earlier. In April of 1993, Michele Chapman was fifteen years old when she and a friend were attacked in a Zachary cemetery. Zachary is a small town of about 12,000 residents and is located in the northern part of East Baton Rouge Parish.

The attacker was a machete-wielding black man who dropped his blade and ran away when a police officer happened upon the "crime in progress." In wasn't until 1999 that Derrick Todd Lee was identified as the attacker. His identity came to light after the case was aired on TV's *America's Most Wanted.* Michele was contacted and picked Lee out of a photo lineup. In the end, the district attorney elected not to pursue the case as there was not enough evidence. Michele ended up with some scars on her ankle as a permanent reminder of her close encounter with death. She was lucky that night because this animal was intent upon killing them. Of course everything is clearer when viewed in hindsight, but now we knew that the killer had a history of attacking women in cemeteries.

LAMBS

Like everyone else, I was trying to make some sense out of all this. The killer had certainly been on a learning curve. The first two DNA-linked killings occurred in the homes of the victims. This in itself raised a litany of questions. *How did he get in? Was it a case of opportunity? Stalking? Did he get into Gina's house when she let her dog out? Was he a smooth-talking con artist who could get someone to open the door by a ruse?* Plus, we didn't really know what motivated his choice of prey. The one thing about him I was sure of was that he was going to keep on killing.

Human response to such things as terrorists and serial killers has not really changed much over the centuries. The Italian writer Giovanni Boccaccio classified the reactions of Florentines to the plague of 1348 into four general responses. We saw that same full gamut here: some women became reclusive and would not venture outside without an escort; some took appropriate precautions; some simply moved away; and some women chose to take no heed to the danger and kept going about their lives, taking chances. I believe these latter were in the minority. But it only takes one.

For instance, I was driving by LSU with my wife when we saw

a pretty coed jogging along a rather secluded stretch. She was by herself! It was around five P.M. And as my grandpa would say, I was flabbergasted! I assumed she lived in one of the LSU apartments several blocks away. I pointed her out and grimly said to my wife, "There goes the next victim." De concurred.

Yet here was this coed bopping down the road with a head-set on. *Idiot!* She would never hear anything coming. Not that it would make much difference. She would be easily overpow-ered. I wanted to get out of the car and confront her. I wanted to ask her if she had been keeping track of current events. What the hell does it take to get your attention? I wanted to show her pictures of the bodies of the victims—pictures that have been seared into my brain. I don't want her picture there.

Security is inconvenient, and we don't like inconvenience. The attendance at self-defense seminars declined. More women were out alone—or maybe I just noticed it more now. You could buy Mace just about anywhere in Baton Rouge, and that was a good thing, but fewer women seemed to be carrying it. I found myself looking at women to see if they had it with them. I suspect the killer was looking, too.

The death of an innocent causes a great deal of outrage. You desperately want to stop this guy before he kills again. He's kill-ing us—not someone who's put herself in harm's way. It be-comes very personal. You are driven to catch him. At the same time, you must look at our value system. Why didn't we do the full-court press for the Black Prostitute Killer? For one thing, they didn't have the strong family advocates that the Baton Rouge serial killer's victims had. Still, my friend Stan would tell me that our society needs a values clarification process. He's right.

It seemed to me that this guy looked for easy prey. The old adage of "Why hunt for a lion when you can kill a lamb?" rang true for this killer. It is not that hard to become a "lioness," but it is inconvenient.

Denial is another phenomenon that makes for easier victims. The disclaimers of false security began to creep into women's conversations. I've heard some of these myself:

"He's just after young attractive women."

"I don't think I fit that profile. I'm not too worried."

"I think he's moved away from Baton Rouge."

"The police have scared him away from here."

They were wrong. Dead wrong!

CHRISTMAS EVE

It was late afternoon on Christmas Eve Day 2002, when sixty-five-year-old Mari Ann Fowler, an attractive lady, and prominent in Louisiana, exited the interstate onto Louisiana Highway 415 in West Baton Rouge Parish. It was the same day that the newspapers reported the serial killer had been DNA-linked to the death of Dené Colomb.

Mari Ann had been traveling west along I-10, the same interstate that crosses over Whiskey Bay about thirty miles west of where she was at that moment. Her initial destination was Lake Charles, Louisiana, and from there she planned to travel on to Texas.

Once she turned onto LA 415, she traveled a half-mile or so to a Subway sandwich shop located in a little strip mall known as the Plaza 415 Shopping Center in Port Allen. At 5:30 P.M. she got out of her car, went inside, and purchased an order to go. She then exited the Subway but never made it back into her car.

It happened at 5:40 P.M., and it happened suddenly—a blitz attack in a populated area, in front of an open shop and in daylight. *How brazen can you get?* The "sandwich specialists" in the Subway didn't see the abduction but did notice that Mari Ann's car was sitting there long after she had made her purchase. Her keys, the sandwiches and her purse were on the pavement beside her car. Wrapped Christmas presents were in the backseat and were undisturbed. Robbery was definitely not the motive for the crime perpetrated here.

There was a surveillance camera in one of the stores in the strip mall, but initially it yielded little information. However, the film was sent off to the FBI for enhancement and analysis. It showed that a violent struggle ensued during the abduction and that the assailant drove a 1994 or 1995 Chevrolet pickup truck that was dark in color.

Mari Ann Fowler happened to be the wife of former Louisiana elections commissioner Jerry Fowler, who had held the statewide office for twenty years, before pleading guilty in 2000 to bribery in a kickback scheme involving voting-machine parts. She was on her way to a holiday visit with her husband, who was serving five years in a federal prison in Beaumont, Texas (he was released in June 2005).

She has still not been found.

Her son, John Pritchard, age forty-two, was very close to his mother and took her disappearance especially hard. Historically they had communicated on a daily basis. John is what we call a "big ole boy" down here. He is a big, gruff-looking man who is gentle as a lamb once you get to know him. His emotional pleas and demands resounded from the State Capitol steps during the rallies and he seemed somewhat lost and overwhelmed by it all. Who can blame him? Not I.

In May of 2004, State District Judge Janis Clark declared that Mari Ann Fowler was dead. I don't know that she was a victim of this same killer. No link with Derrick Lee has been established to date. But I do have a high suspicion, as do many other people, that it was him. Lee was driving a 1994 Chevy pickup at that time and it was maroon in color. Lee's cell phone records also indicate that he was within several miles of the abduction site on that day. Mari Ann fit his victim profile. You can take it from there. I have. But I can't prove it and he is innocent of her death until proven guilty.

CARRIE LYNN YODER

Public fears climbed steadily. In January 2003, the task force announced a major development in the Baton Rouge killer case. They found another shoeprint, allegedly from the killer, in the vicinity of Dené Colomb's body. It was the second shoeprint to be found in connection with the killer. The first was found at the murder scene of Charlotte Murray Pace. The print was that of an Adidas sports shoe, men's size 10 to 11. It's a pretty common shoe, and pictures of a similar shoe were posted all over the TV and on the task force website. I doubt the killer kept it around after all that. Still, it was a clue. But the killer was still considered to be a white male. More important perhaps is the fact that the shoeprint clue, as well as the pressure from the victims' families, kept the killer in the forefront of everyone's minds.

The families of the victims were the conscience of the community. There were those who supported their actions and those who criticized them. I think of the critics as those who cast the proverbial first stone. I have never lost a child to a killer. I cannot stand in their shoes. I can only empathize with them and know that at the end of the day I have done my part to the best of my ability to catch a killer.

I am responsible for the dead but I am responsible to the living. I do know the sadness, if not agony, of talking to family members when you have so little to say. The families kept these murders in the spotlight; and community awareness would ultimately lead to the killer's arrest.

Four months after Colomb's murder, he struck again. On March 3, 2003, twenty-six-year-old Carrie Lynn Yoder, an ecology doctoral student at LSU, went missing from her home on Dodson Avenue—a stone's throw from LSU. She lived only a few miles from Gina Green and Charlotte Murray Pace. Like Colomb in Lafayette, she lived alone. Her boyfriend of three years, Lee Stanton, reported her missing to police two days after she failed to contact him. They had returned the day before,

Sunday, from a trip to Mardi Gras in New Orleans. She was last seen wearing a dark sweater-jacket and blue jeans, and her long, dark, wavy hair was pulled back. The groceries she'd told her boyfriend she would buy were on the table, and her two cats—Toby and Nina—were unfed.

"I feel cheated that the last conversation I had with her was about the groceries," Lee told the press. He *was* cheated.

She was considered a missing person by the police and efforts were launched to locate her. My oldest son, Christopher, is a movie buff and he remarked to me that this seemed like something out of the "vampire flick" *Lost Boys*; the person goes missing, the posters of their face get plastered everywhere, flashed on every TV screen, and then they are found dead and mutilated, and everyone is powerless to stop the horror. The analogy was apt, only this monster was real and so were the victims.

Lots of questions go through your head when a young girl who fits a serial killer's victim profile is missing. Is it him? Is it someone else? Is she alive? Is he keeping her alive for his own iniquitous reasons? And as long as she is missing there is that spark of hope that she will return.

Ten days later, her body was found by a fisherman in Whiskey Bay, about thirty miles from the city and just a few hundred yards from where Pam's body had been found. As we had previously ironed out any cross jurisdictional issues, I went there myself and retrieved her body from the water. In turn, I would keep the local coroner, Dr. Freeman, apprised of my findings.

The scene was controlled. But there was one hell of a crowd

there. There were all sorts of media trucks and eye-in-the-sky cameras. Once I navigated through the media jungle and I arrived at the edge of the water, I got onto a waiting flatboat. The coldness of the metal made the short trip seem even more ominous. I was escorted out to the site a little too fast for my taste, as the flatboat had a very slippery deck; and besides, we didn't need to create a wake near the victim's body. Her body was face down in the water between the eastbound and westbound lanes of I-10, which rise about twenty feet above the water's surface. At first all I could really see was some sort of tattoo on her lower back. One of the detectives told me Carrie reportedly had a similar tattoo. No one spoke much for a while after we heard that.

I employed a special mesh net body bag to collect her remains and any evidence that might be adhering to it. As a side note, the mesh was so effective that we actually retrieved one of her loose contact lenses. It was very difficult working in the water as we were trying to minimize any disturbance of the body. In the process, I managed to drop my cell phone into the shallows—*brilliant move, Lou!* While retrieving my phone, I noticed the water actually felt cold, and there was minimal if any minnow activity in the shallows. There was minimal or no animal activity on her, either. I wondered about crawfish and turtles and catfish and whatever else lived in this swamp. Where were they and why didn't they get to her? These questions would indeed require scientific investigation because we needed to know how long she had been there.

Once Carrie was secured in a body bag and all evidence

sealed in place, she was placed into the coroner van and one of my investigators braved the amassed crowds with police help and escaped onto the interstate, headed east. That crowd would relocate to the morgue in short order.

The trailer morgue is located in the city parish garage parking lot. It actually stands near the vehicle washing bay—away from any occupied structures. It is encircled by a chain-link fence, which allows for control of entry and security. That is one of the reasons I chose the location, and it has proven to be a good place for our morgue.

Heavy window coverings were put in place to discourage prying eyes and camera lenses. The main reason that we performed her autopsy that same day was to maximize the amount of evidence that we could glean from the procedure. She had been strangled, but not before being savagely beaten. While she was still alive, nine of her ribs were broken at the point where they connect with the spine. The broken bones punctured a lung and her liver. We knew she had been alive at the time of the beating because of the hemorrhaging at the trauma sites. You don't actively bleed when you're dead. A forensic autopsy is a reconstruction of the immediate events that led to a person's death. As such, you relive the events with that person. People tend to look at you like you're crazy when you say you are "listening" to the deceased. It's an intensely respectful process, and neither Dr. Cramer nor I indulge in or tolerate any of the gallows humor that is allowed in some other autopsy theaters. I don't want to hear the excuse that such humor or inappropriate remarks are a way of dealing with the stress of the situation. It's inappropriate—period. If you do it in our morgue you are going to get your feelings hurt because we're going to invite you to leave, and we're going to do it publicly.

As anticipated, a veritable hoard of reporters had followed us from the scene, requiring the police to cordon off the area near our little trailer. The autopsy was a long, arduous process, and I left there at around one in the A.M. Unbeknownst to me, some reporters had managed to slip through the police barrier and as I began to walk to my car, several descended upon me. I'm afraid I was a little short with them. The extremely bright camera light that flooded my face, nearly searing the rods and cones in my retinas, didn't help my mood any. I got on the horn and had the police come remove them. I was a little more pissed off at the local guys who did that because I have always been open with the press—those I trust, anyway. I guess they, too, were caught up in the media feeding frenzy. It seemed to be a time of excitement for them, when it should have been one of alarm and mourning.

Ultimately, I had one of my investigators stay in the morgue trailer all night, for fear some reporter might try to gain illegal entry into the facility. What a macabre circus.

The next day, I was on the phone to meteorologist Jay Grimes, because I needed information about the weather in the area for the past ten days. I needed to know the temperature, the cloud cover, the amount of precipitation, and anything else he could tell me. These are all factors that can change a crime scene; they are also factors that influence insect activity and the rate of body decomposition. Jay is one of the good guys who always steps forward, helps in any way he can, and asks for nothing in return. This time was no exception.

One of the haunting questions in everyone's mind was the possibility that the killer had abducted her and kept her alive for a week or so. It fell to me to answer that question.

I also touched base with Wildlife and Fisheries about aquatic animal activity in the area. Evidently there is a sizable layer of silt on the bottom of the bay, and if she had been thrown from a height, as we surmised, it is quite possible that she was actually submerged in the silt. When the gases in her body formed and expanded due to the normal putrefaction processes, her body would have risen up out of the silt to the surface.

I also contacted Dr. Bill Bass, head of the Forensic Anthropology Center at the University of Tennessee, in Knoxville, widely known in forensic circles as "The Body Farm," the only facility of its kind in the United States that allows students to study bodies in various states of decomposition. I related the condition of the body and the ambient conditions to him. Bass, a renowned forensic expert whom I knew via a mutual friend, has assisted in hundreds of death investigations, from historic homicides to large disasters.

After exchanging views, we agreed that the condition of her corpse was compatible with her being in the water for a week to ten days—roughly the same period she'd been missing. We concluded, as a result, that he had not kept her. She was abducted, beaten, raped, and strangled, then dumped into Whiskey Bay from the interstate above. That's at least a twenty-foot drop. What kind of abhorrent animal could do this? I certainly didn't know it at the time but we were close to finding out. But not close enough to have saved this young woman's life.

More than anyone, the parents suffer in such deaths. Lynda and Dave Yoder, both of Florida, learned of Carrie's disappearance on March 5th. Dave called his wife at the end of the workday to say he'd be home with "news of a problem." *Can you imagine going home to tell your wife that your child is missing? I truly cannot grasp the degree of emotional pain he must have carried with him.* They drove ten hours straight to Baton Rouge that night. Their only stop was in Gainesville, Florida, to see their son, Greg. As Lynda Yoder said: "I just had to hug my son."

The day that Carrie's body was discovered was also Dave's fifty-eighth birthday.

I met Carrie Lynn Yoder's family at Rabenhorst Funeral Home the day after her autopsy. We were in one of their quiet, tastefully furnished offices. The door was closed for privacy. I felt so inadequate. The room was heavy with the pain and grief that accompany losing a loved one in such tragic, inexplicable circumstances. I was there to give them the jewelry we had taken from their daughter's body and to answer any questions. They had none. I think they were still in a state of shock and did not want any details. One of the things that hit me was the sense of how truly united they were as a family.

From the recesses of my mind came my grandma's words about real family, as she called it: "Sometimes when you see people at their worst, you see them at their best." I now knew what my grandmother meant.

There was also a Pandorian cornucopia of negative emotions ricocheting through my mind. Some of those relate to guilt and embarrassment for myself, and for Baton Rouge. These people had sent their daughter to our city, to study at our university. They had trusted us to take care of her. To provide a safe city. We had let them down, and now they were here to take the dead, abused body of their daughter home.

Could I remain objective and do my job? Yes!

Could I compartmentalize the emotions and focus on the science. Yes!

Did I want this guy to pay for Carrie and the other innocents? YES!

And I wanted him to pay dearly.

On March 18, 2003, two weeks after Carrie was reported missing, and about four days after we discovered her body, authorities announced the DNA link of this murder to the serial killer. Carrie's was the fifth murder linked by DNA since September 2001, following Gina Wilson Green, Charlotte Murray Pace, Pam Kinamore, and Trineisha Dené Colomb.

At the press conference, Governor Mike Foster appeared with representatives from the FBI, the U.S. Attorney, the Multi-Agency Homicide Task Force, a state representative, and Baton Rouge mayor Bobby Simpson. Each vowed justice. "Money won't be an object, personnel won't be an object," Foster said.

What frustrated a lot of people was that the suspect left no "signature." The media asked me repeatedly about this. I never did quite understand that mindset. I think this whole "signature" thing was a contaminated expectation—contaminated by the many fictional accounts of our line of work on TV. What more signature do you need than his DNA? As the press pointed out, Green and Yoder were strangled, Pace and Kinamore's throats were slashed, and Colomb was beaten to death. Victims were white and black. They ranged in age from twenty-two to forty-four.

There were, however, similarities: no crime scene showed signs of forced entry, which hinted that perhaps the killer was able to talk his way into victim's homes. All of the women were sexually assaulted. And the killer never stole anything of great value from the homes—though three phones were missing, and there was that phone cord at Whiskey Bay. Several people voiced legitimate criticism about the vagueness of the FBI profile, while so-called experts went on news shows, droning on about the psychological profile of this monster. In my world, armchair speculation is cheap, and cookie-cutter investigations are worth even less. There would be no magical solution.

THE BELL

I cannot imagine the anguish of losing a child. I simply cannot. I don't believe anyone can, with the exception a parent who has experienced that loss.

There are no adequate words to soften their sorrow and grief; at least none that I possess. I've seen the devastating consequences of such a loss. Disbelief, pain, and turmoil are often the harsh companions of grief.

When that loss is the result of a vicious unprovoked murder, the grief seems magnified, if that is possible. Professionals in the field of grief and loss tell me that grief may become prolonged due to lack of resolution at certain levels. One such "level" is not having the killer brought to justice. I'm not a grief therapist but I figured that one out on my own.

De and I were standing on the steps of the State Capitol in the heat of that fireball we call the Louisiana sun on March 16, 2003. It was an especially hot day. There is no shade on those steps. There is only pale stone and concrete. Both are unforgivingly reflective. Umbrellas were out and opened to block the merciless heat and glare. We hadn't had the foresight to bring one, of course. Water was being offered to prevent dehydration, and we accepted it gratefully.

There was an element of grim determination among the crowd of several hundred who had congregated there for a memorial service for the victims of unsolved murders in the parish. It was organized by some of the family members of the women murdered by the serial killer still at large in the area. It was more than just a memorial service. It was also a plea for women in the area to be on guard and to not become the next victim. It was fueled by the determination that their daughters' killer would be brought to justice.

The family members' demands and criticisms were not received with open arms by all involved in the case. Many said the task force wasn't on the ball; Pam Kinamore's brother-in-law called local politicians "wimps" and urged clergy and businesses to throw their weight behind the investigation.

Lynne Marino, Pam Kinamore's mother, read a poem that ended with two lines that she could speak only through tears:

Do you know what it's like to bury your child?
I do, and I hope you never have to know what it's like.

I'm not sure what my expectations were. I had anticipated a range of expressions of grief on the part of the family members. I understand rage and fear. I understand the need for answers, even when none are forthcoming. I understand the need to attach blame in the face of innocents being horrifically murdered. I understand tears, and the courage to go on. I understand my wife wiping the tears from her eyes. I know about this thing, grief, and I am ready for it. It happens. It's supposed to happen in circumstances like these. I understand it and I anticipate it.

I did not anticipate the bell. I was not ready for *that*.

Lynne Marino made the announcement: "We are going to ring the bell once for each female murdered in East Baton Rouge Parish. These are women whose murders have not been solved. We will call out their names as we ring the bell."

I've come to know Pam's family. They demanded justice for Pam and for the other victims. Pam's mother is one of the most courageous women I've ever had the honor to meet. Thank

God for people like her. They keep us accountable, on many levels.

I'm a very visual person, and I tend to retain images in my memory. I don't know if that is a blessing or a curse. Today it felt like a curse. When the bell tolled for a victim whose murder occurred on my watch, a vivid visual image of the dead woman flooded my brain. It was a horror show—one after another . . .

By now I was oblivious to the heat and the sweat. I felt like it was Judgment Day and I was being held accountable by these images of the cases I had worked. *Did I do my best? Did I miss anything on my end? How did I fail? What am I doing about it now?* One after another, the names continued, and so did the pictures. It seemed endless. I'm ashamed to admit that I was glad when the service was over. But it will never be over for those families. They exemplify the love and dedication family members share with each other.

At another level, I humbly submit that I don't think it will ever be over for me, either. Am I comparing my own emotions to theirs? Certainly not. I cannot imagine the abyss of pain they must feel. My point is this: dealing with unexplained violent death and trying to be a part of bringing justice to the victims changes a person forever.

I'm thankful I was there that day. I'm thankful for the families and the others who came forward. It was one of those pivotal experiences that made me take stock of myself and my profession. That day I became even more convinced that it would take a community effort to stop this killer. It's a sickening feeling to look around and wonder if one of the women here would be the next victim. By increasing awareness among women, we could decrease the vulnerability of potential victims. I vowed to do just that.

These were times of turmoil, but I knew my charge. *The coroner is responsible for the dead and to the living.* That was the first memorial rally we attended. It would not be the last.

FOURTEEN

To Catch a Killer

CLOSING IN

On March 21, 2003, the Multi-Agency Homicide Task Force announced in a press conference that the Louisiana serial killer could be of any racial background—this after suggesting for eight months that the person who committed five known murders was a white man driving a white pickup truck. True, the task force had warned from the start that people should not screen tips based on race; but that did nothing to quell widespread speculation that the suspect was Caucasian. Pam's assailant was perceived to be a white man in a white truck. Not only that, he was a white man with a medium to thin build. The perception that the killer was white was reinforced when the

Lafayette Parish Sheriff's Office released its sketch of a "person of interest" who had been seen near the woods where Dené Colomb was found. It was an image of a white guy.

This new "race" announcement caused a great deal of consternation for many people, but none more so than the Sharlo witnesses who, having provided a composite sketch of a black male as their suspect, found their reports to be repeatedly discounted by the police. Not a good thing!

Tony Frudakis, chief officer of DNAPrint Genomics, had changed the focus of the investigation. In February 2003, he actively pursued the Louisiana State Police Crime Lab to use his technology to determine the killer's ethnic background.

The task force sent twenty DNA samples to his lab. Nineteen of the twenty samples came from known donors. The twentieth came from the killer. The results indicated the killer's ancestry to be 85 percent sub-Saharan African and 15 percent American Indian. Frudakis says he relayed this information to the task force in the first week of March. They had been focusing on the wrong guys. The killer was a black male!

On May 26, 2003, in the largest headline I have ever seen in the *Baton Rouge Advocate,* the word "WANTED" was printed above the face of a thirty-five-year-old black male named Derrick Todd Lee, who had just become the target of a national manhunt. Described as six-one, about 210 pounds, with a light to medium complexion, short hair, and a muscular build, he had a history of arrests and convictions, including peeping tom violations, burglary, and attempted first-degree murder. He was a cement-finisher and truck driver from St. Francisville, a quaint West Feliciana Parish town located north of Baton Rouge. The little town is known for its plantation homes, bed-and-breakfasts, and antique shops—not murderers and rapists.

The task force had repeatedly spoken of its reliance on the community to provide information that would solve the case and had also repeatedly talked of getting a lucky break. The "break," if you want to call it that, came in the form of a citizen's tip from Zachary, Louisiana. And it centered on a murder that the task force had not previously included in its purview.

Five years earlier, on April 19, 1998, a neighbor found the three-year-old son of Randi C. Mebruer playing alone in front of his house in Oak Shadows subdivision. When asked, the boy simply told the neighbor that his mommy was "lost." The neighbor ventured inside the home in search of the child's mother but found only signs of a violent struggle and a great deal of blood throughout the house. She called the cops. Randi, a twenty-eight-year-old white female and home health nurse, had been abducted from her home.

She lived about one block from Connie Warner, a forty-one-year-old white female who had been abducted in August of 1992. Connie's body was found in a ditch two weeks after the abduction. There were no signs of forced entry in either crime, and both were undeniably similar to the abduction and murder of Pam Kinamore and Carrie Lynn Yoder.

Police officers in the area were quite aware of Derrick Todd Lee's peeping tom history and he was immediately placed on the short list as a suspect. The day after Randi was noted missing, police interviewed Lee at his home on Highway 61 in St.

Francisville. He consented to a brief search of his premises before asking them to leave. They did—with him in tow. He was interrogated at the East Feliciana Parish Jail, but of course denied any wrongdoing.

His story was straightforward and confirmed by his girlfriend. On the night that Randi was abducted, Lee was partying with his girlfriend at the Highland Bar there in St. Francisville. Lee's flirtatious advances toward another woman in the bar were not appreciated by his girlfriend and a spat ensued between her and Lee. She bailed out on him at about 10:30 P.M. He subsequently ended up at the Hideaway Bar in Alsen, Louisiana. He drank there a while, then headed over to his girlfriend's house in Jackson, Louisiana. His route would have taken him right by the entrance to Oak Shadows—Randi's subdivision.

The other prime suspect was, of course, Randi's ex-husband, Michael Mebruer. He complains that he was followed and investigated by the authorities for years and still harbors ill feelings toward his pursuers, particularly one Dannie Mixon of the attorney general's office.

It was generally thought that the Baton Rouge Serial Killer, who had since become known as the South Louisiana Serial Killer, never spoke to anyone about his grisly conquests. But he did talk about Randi Mebruer. In fact, the arrogant slime-ball bragged about it late one night to an acquaintance. His words were "They will never convict me, because they will never find the body." That boast would be his undoing, albeit five years and numerous homicides later. The acquaintance worked on the property of James Odom and subsequently relayed the conversation to Odom, who relayed it to the police. Again, all this occurred five years before Lee was charged with any murder.

Odom left matters with the police, but he never forgot those haunting words. And they came back to him as the murders in Baton Rouge continued. But the serial killer was supposed to be white, and Lee was black, so Odom was thrown off some. He did, however, discuss the matter with his son, Joel, who is a detective with the East Feliciana Sheriff's Office. Joel in turn discussed the case with the Zachary police, who, in his view, didn't take it any further. But he didn't let go, and at the beginning of May, Joel Odom met with Dannie Mixon, to whom he relayed the information from his father. Mixon followed up quickly with an interview of the workhand.

Mixon, a sixty-five-year-old investigator of the "old school," had worked in the attorney general's office for twenty-two years. A local boy, he was a graduate of Baton Rouge High School and LSU. He had worked in law enforcement all his life, having joined the Louisiana State Police when he was a seventeen-year-old college freshman.

Mixon had assisted Zachary police with the investigation into the disappearance of Randi Mebruer. His focus had indeed been on Michael Mebruer, but what he heard now from Detective Joel Odom, and from James Odom's hired hand, made him change his mind.

From his investigation of Lee's criminal past, Mixon knew that Lee was not in prison at the time of the killings, and that he was a known burglar and peeping tom. Both of these crimes tend to go hand in hand with sex offenses. Lee also had a significant history of violent behavior. In January 2000, Lee was booked on aggravated battery, attempted first-degree murder, and aggravated flight from an officer. He was accused of beating his girlfriend in the Liz Lounge, in Solitude, Louisiana. It was a recurrent theme: he flirted with another woman, she didn't like it, so he tried to stomp her to death while wearing cowboy boots. Then he tried to run over a sheriff's deputy when he fled.

In April, all but one charge was plea-bargained away and he was convicted of flight from an officer in West Feliciana Parish and sentenced to prison for two years. A judge also revoked his probation on a previous stalking charge, and ordered him to serve nine months. Here the legal system had failed us. I mean, let me get this straight: He breaks probation for stalking, he stomps a woman and tries to kill a police officer, and he only gets two years for it? Two years for two attempted murders, one of a police officer! It seems unbelievable, but that is exactly what happened. In January 2001, he was officially discharged from the Louisiana Department of Public Safety and Corrections, having served only one year. *No wonder this creep thought he was invincible and above the law!*

Mixon determined that Lee was the most likely candidate out there, and District Judge George Ware agreed, signing a subpoena to obtain Lee's DNA. It stated: "Lee has remained a viable suspect in the disappearance and alleged death of Mebruer and murder of Warner and a possible suspect in the deaths of five females in the Baton Rouge and Lafayette, Louisiana, areas that have been linked by DNA profiling to the 'serial killer' operating in the Baton Rouge and Lafayette areas." The door was closing on the killer.

On May 5, 2003, Mixon set out with local police officers to do a DNA swab on Lee. Lee wasn't too happy about being approached. He knew who these cops were and tried to evade them in his vehicle, but to no avail. When he was finally cornered like a rat, he cursed the officers for trying to frame him. *I'd say that qualifies as suspicious behavior!* Lee, however, calmed down after he was swabbed—by Mixon himself—and reportedly became as docile as a "whipped puppy."

After the swabbing, Lee remained in the area. Indeed, two weeks after the incident, Police Chief Kenny Simmons actually "bumped into" Lee walking down the street in St. Francisville. Lee even initiated a conversation with the chief and shook his hand! *I can't really comprehend this. With the same hand he beats and strangles women he shakes the hand of the police chief! Good versus evil on the streets of St. Francisville. St. Francisville for God's sake! It's a Mayberry look-alike where I half-expect to see Andy and Opie walking around—not this. I guess it all comes back to Stan's explanation. Whaddya expect, Lou? He's a sociopath!*

And where was the DNA swab while Lee was casually chatting with the police chief? It was still sitting in the State Police Crime Lab, waiting to be processed!

It was about this time that Louisiana authorities caught up with Lee. St. Martin Parish detectives released the sketch of a man resembling Lee in connection with an attempted rape in Breaux Bridge, and on May 23 Chief Englade of the task force expressed the belief that the black male in the sketch might be the killer they were searching for.

Finally it was all coming together.

The would-be rapist had talked his way into a home by asking to borrow a phone and phone book. His attack was interrupted by the woman's son and he fled the scene in a gold 1997 Mitsubishi that had a dent in the hood and a "Hampton Has It" license-plate cover. Lee had a 1997 Mitsubishi registered in his name; as a matter of fact it was sitting in his yard, complete with dented hood and license-plate cover.

On May 24, the sketch was plastered on the front page of the newspaper. The resemblance to Lee was unmistakable. Zachary police recognized the face and asked the crime lab to do a rush on the DNA submitted from Lee. On May 27, the sketch was replaced by a photograph of Derrick Todd Lee, because the DNA results were back—after three weeks—and he was identified as the South Louisiana Serial Killer. But by then, Lee had fled the state. *I guess he figured that unlawful flight to evade prosecution was the least of his concerns.*

A nationwide manhunt was launched. But Lee didn't seem too upset by it all. At least that's the impression he gave to fellow residents at the Lakewood Motor Lodge in Atlanta, Georgia. He'd checked into the cheap motel and rapidly made friends. He enjoyed partying and was evidently pretty good at grilling ribs and chicken for the guests. He was even known to "entertain" women is his room on several occasions during his ten or so days in the area. *I wonder if they knew that they were being "entertained" by a man wanted for murder?*

On the night of May 27, 2003, Derrick Todd Lee, the monster who had stalked our streets for so long, was arrested without incident outside a tire shop in Atlanta. The police asked for his ID and he turned it over to them. That was it. Lee's days of raping and killing were over. But an aftermath was dawning; and it wasn't going to be pretty.

CRITIQUING THE CRITIQUE

I thought about avoiding this topic altogether but I feel I must weigh in on the performance of the Multi-Agency Homicide Task Force in order to get some closure for myself and others involved in this whole tragedy. Let me first state that I embrace the concept that if we do not learn from our mistakes we are destined to repeat them.

For weeks, the press hammered on the task force for not having made the critical connections sooner. Many felt that if the task force had done a better job, the murderer would have been caught earlier and the killings would have stopped sooner. The day after Lee's arrest, members of the task force who had been on the case for eighteen months defended their strategy in a combative, sometimes tearful news conference packed with reporters, law enforcement, and supporters. Baton Rouge police chief Pat Englade said that by the end of the investigation, the task force had followed up on 26,000 leads and swabbed more than 1,200 men. Englade was emotional when he said he'd have wanted the very same investigators if his own family were involved in a crime. "They have been totally, unduly criticized," Mayor Bobby Simpson told listeners. "We owe them a debt of gratitude that can never be repaid." With Governor Mike Foster at his side, he called them "one of the most competent task forces that has ever been put together."

Maybe so, maybe so . . . but if we apply the same yardstick to criminal-justice professionals as we do to medical professionals, we'd have to come back to the astute words of my med-school mentor: "The operation was a beautiful success! Too bad the patient died." It's a fitting analogy. Like I said before, in our profession, 99 percent is a failing grade, and we know that going into it.

Could they have done a better job? Only they can answer that. I know some of those guys; they're good guys, and I am sure they have done, and continue to do, a lot of soul searching about that very question. There were definitely some problem areas. In addition to not communicating well with or through the media, they didn't communicate effectively with fellow agencies. I'm not even sure how well they communicated within their own agencies. They set up operations in a building so far removed from police headquarters that their own street officers began to refer to it as "the ivory tower," or "the bunker." The heavy reliance on DNA is perceived by many to have taken the place of basic police work. This opinion was magnified when Mixon, Odom, and the Zachary PD were credited with breaking the case. They were not on the task force though. An offer of help from the attorney general's office had also been declined.

One rationale at the time for the public relations fiasco was that the task force chose to keep facts that only the killer would know away from the public. This approach allows investigators to positively confirm anything a suspect might incriminate himself with during questioning. I understand that. But at the same time there is the need to keep the public informed and to treat family members with the care and respect they deserve. Their public relations effort was disastrous at times, and that cost them heavily.

Of course, with the benefit of hindsight, any critic can look brilliant. The press created a platform for critics, who charged that the task force missed or even ignored clues, and as the inevitable deconstruction began in the media, it was clear that the task force was in for a bumpy ride.

Did the task force fail to communicate effectively with nearby forces? Robert Keppel, a lead investigator in the Ted Bundy murders, pointed out to the local press that the task force should have looked to surrounding parishes to see if similar crimes had been committed and relevant suspects studied. I think we already knew that. It's called Investigation Techniques 101. I also think we have established the fact that agencies tend not to communicate much with each other. But in this case, there are some exceptions. The Zachary police said that they gave information about leading suspects, including Lee, to task force officials in summer 2002 at a round table session with other law enforcement in attendance. They also presented the facts that he was a convicted peeping tom and that he was a suspect in the only two unsolved murders in Zachary. If Lee had been swabbed then and the DNA analyzed, Dené and Carrie might be alive today. That's a harsh reality.

St. Martin Parish said they informed the task force of the Breaux Bridge attack shortly after the task force formed in August 2002. Englade said he didn't hear about it until April 2,

2003. If that's true, the glaring question for the chief is: "Why not?"

Did the task force rely too heavily on DNA analysis to catch the killer? Chief Englade told reporters that his agency had not linked the serial killer to the Zachary crimes because the Zachary cases hadn't been analyzed for DNA. "We drew a line in the sand because of DNA evidence that we would only link the ones that we have DNA evidence for," Englade told the Associated Press. "But it's not that we were overlooking anything else." *Huh? I'm confused. What exactly does that mean?*

Some critics lay blame on the state itself. Perhaps Lee could have been caught sooner if the Louisiana legislature hadn't delayed, by nearly two full years, funding for a state-mandated program to perform DNA sampling of arrested suspects. Unfunded mandates are little more than feel-good legislations, as far as I'm concerned. Still, I'm not sure how much difference it would have made. The vast majority of DNA samples collected by the task force were from white males.

Did the task force listen to the public? In 1999, Derrick Todd Lee pushed his way into the apartment of Collette Dwyer. His stated purpose for the intrusion was to "take care of her." This was the culmination of his having stalked her for some time. He ultimately was arrested for stalking, pleaded guilty, and was placed on probation. The stalking resumed in 2001. Collette called the tip line to report Lee but says she was told that they were looking for a white man. Her call has been verified and confirmed. Collette placed her first tip about Lee after the death of Charlotte Murray Pace. She called authorities in Baton Rouge again after the murder of Pam Kinamore. She also told authorities that Lee drove a white pickup. Evidently an investigator from the task force went out to look at the truck but did not think it appropriate to swab Lee, due to his race.

After the murder of Charlotte Murray Pace, a police detective said later during the trial, four people told police they saw a suspect similar in description to Lee near Pace's house on the day she died. "They gave general descriptions," the detective said, but accounts differed. When police refused to prepare a composite sketch, the neighbors hired a criminologist to prepare their own. The task force said it would not release the sketch because "We have no proof he [i.e., the man described by Pace's neighbors] had any connection to . . . Pace's homicide. . . . We don't know why he was in the area." That really sounds pretty flimsy to me; especially when the police would go out and swab someone in the blink of eye over an anonymous tip or declare someone else a person of interest simply because they were seen in a truck near a crime site. I mean, come on— shouldn't the guy at least have rated a person-of-interest status?

Did the task force provide an accurate profile or description of the killer? The FBI profile was on target in some areas but vague in others. But aren't they all? It's just a tool. A profile is also a work-in-progress that tends to change as new data come in. The profile was right about Lee being physically strong and a low wage earner. It also correctly predicted his age to be between twenty-five and thirty-five. One glaring misstep was that the profile labeled him as being "probably awkward around women." That was an unfortunate statement in that it was so very wrong. Again, there are no absolutes, and the profile bore that disclaimer. But people have selective reading, just as they have selective hearing, and disclaimers are easily perceived as perfunctory in cases like this.

If a woman read and believed that the killer was an awkward white male, then a black male who was charming around women might readily be excluded as a threat. The woman excluding him could also become his next victim. He was looking for lambs who were easily disarmed. *In effect, the profile was helping to disarm them!* Those who knew Lee describe him most often as gregarious, charming, and flirtatious—not awkward. When he was caught in Atlanta, he was luring women to his room with the promise of cognac. Of course, he was also leading a Bible study group between barbeques.

Even though task force officials met with families of the victims, that's exactly where some of the most heated criticism, understandably, came from. At one of several rallies held on the capitol steps, parents, husbands, siblings, and friends vented their mounting frustration on local authorities, whom Pam Kinamore's brother said were "in over their heads." Pam's mother, Lynne Marino, appeared on TV to air her frustrations at the investigation. When police finally captured Lee, Marino, in a barb at investigators of the Multi-Agency Homicide Task Force, told the *Advocate*: "Well, I'm thrilled. I was worried that he was going to turn around and come back to his safe haven, Louisiana."

Most of the comments came from a place of deep pain and loss. Carrie Yoder's boyfriend said he was relieved at Lee's capture, but also confessed, "I'm very, very angry. I have so much rage inside me right now that I feel like I'm coming apart at the seams."

I am thoroughly convinced now that there is no ultimate authority on serial killers. There is no cookie-cutter solution in the face of this type of madness. No one has all the answers.

In the end, I sent Chief Englade an email of congratulations on the night Lee was captured. I meant it when I said: "You and law enforcement are to be commended for the long hours and efforts that have broken this case." Of course by "law enforcement" I also meant the attorney general and the Zachary Police Department, who were not part of the task force. Indeed, I do not want to lose track of the fact that the evidence the task force assimilated on Charlotte Murray Pace's death, as well as the others, was utilized to convict Lee and get him the death sentence he so deserves. They did a lot of hard work. On the other hand, in my humble, non–law enforcement opinion, their exclusionary strategy proved to be a fatal flaw.

A SURVIVOR'S ACCOUNT

We now have an eyewitness account of the methods Lee used to subdue his victims. That account comes from a very brave woman who survived one of his assaults and lived to confront him face-to-face—in court. Diane Alexander, a nurse from Breaux Bridge, Louisiana, gave her testimony in January 2004, on the second day of Lee's pretrial hearing in the murder case of Charlotte Murray Pace.

On July 9, 2002, three days before Pam Kinamore would be

abducted from her home, a young clean-cut black man who identified himself only as "Anthony" presented himself at Diane's door. He seemed harmless enough, standing there on her porch, claiming he was lost and trying to find "the Montgomerys," friends of his who lived in the area. He requested the use of her telephone. He also asked about Diane's husband, and upon learning that she was alone, he barged into the house. She tried to close the door, but the muscular Lee quickly overpowered her.

Then, in a blitz attack, he grabbed her by the throat, dragged her inside, and pinned her against the door. He produced a knife and threatened to stab her in the eye. His level of excitement was readily betrayed by his profuse sweating. He then pushed her skirt up and tried to rape her. Diane stated that he could not get an erection. By this time, Lee's victim was confused and probably impaired from the lack of blood flow to her brain from the chokehold he had on her. He left her for a moment, to return with a phone cord, which he wrapped about her neck and began to strangle her. Death was imminent.

But Lee was interrupted when a car suddenly crunched its way into the gravel driveway. It was Diane's son. Lee had to bail, but not before one last savage assault: he stomped his half-conscious victim prior to making his escape.

Diane's son, Herman, had missed Lee, but he had caught a glimpse of Lee's car.

Diane was hospitalized with a skull fracture and other blunt-force injuries. Lee's whisper was still in her ears: "I've been watching you."

Lee took the phone cord with him. It was collected as evidence at Whiskey Bay, 450 feet from the recently discovered body of Pam Kinamore.

The high point of her testimony came when Diane Alexander pointed a finger at Lee and said that she was absolutely sure, without a shadow of a doubt, that Lee was her assailant on that fateful day.

It was what the families of the murdered women wanted to hear. They applauded so loudly that Judge Richard Anderson had to issue a warning about such outbursts.

CONVICTION NUMBER ONE

In August of 2004, Derrick Todd Lee faced charges for the murder of Geralyn DeSoto. He pleaded not guilty before the 18th Judicial Court, in West Baton Rouge. As the case unfolded, it was apparent that Lee had evidently resorted to his usual ruse in order to subdue Geralyn. He showed up at the door of her mobile home and asked to use her phone. He even made a call to a number he knew at Exxon.

She was hit so hard that she sustained brain trauma, but she was a fighter, just like Charlotte Murray Pace. But just as with Pace, he stabbed Geralyn to death and slashed her throat. He stabbed her twice in her back and twice near her breast. A reconstruction of the scene indicated that Geralyn had tried to make a dash for her bedroom, where she had a shotgun.

Lee was linked to her death by the DNA under her fingernails. It was his. There was no rape kit, as one had not been done. But the prosecution had Lee in the area, Lee's bloody bootprint, Lee's knife, which was compatible with Geralyn's stab wounds, and Lee's DNA. The jury came back 11–0 on a guilty verdict of second-degree murder. I say 11–0, in that they had eleven guilty votes and one juror simply did not vote. That came as a surprise but then again, only ten guilty votes were needed to convict Lee.

I think it's significant to point out the fact that Amanda Landry, the nonvoting juror, did not vote that Lee was not guilty. She did not vote at all. She reported that she "needed one more thing" to convict him, but she didn't know what that one thing was. She also stated that "they didn't need my vote." That seemed odd to lots of folks I know. The evidence seemed overwhelming, to me, but it's important to never count those chicks before they hatch. I think Landry's most profound remarks centered around the desire she had expressed of always wanting to be on a jury. That statement was quickly followed by "Never again." I guess it's possible for those criminal justice TV dramas to give some false expectations to jurors also. I'm not suggesting this happened in Amanda Landry's situation. But one should be careful of what they wish for.

In August of 2004, Lee was sentenced to life without parole.

CONVICTION NUMBER TWO

Two months later, on October 14, 2004, it took a Baton Rouge jury less than eighty minutes to find Lee guilty of first-degree murder in the death of Charlotte Murray Pace. The defense rested its case without calling a single witness. It was the second murder conviction for thirty-five-year-old Lee. Both Pace and DeSoto were in graduate school, both women were stabbed, and both had telephones taken from their homes. During the Pace trial, prosecutors presented DNA evidence linking Lee to five other raped and murdered women in Louisiana.

The jury rejected claims by defense attorneys that the twice-convicted killer was mentally retarded and thus ineligible for the death penalty. In December 2004, it took them only about ninety minutes to decide on a death sentence by lethal injection.

The pain of the survivors is unimaginable. Most who heard the survivors' statements during the trial could not listen without sobbing. During the Charlotte Murray Pace murder trial, her father, Casey Pace, testified that he was "shocked a heart can keep beating when it hurts that bad." Her mother, Ann, told jurors, "There's no peace in sleep. No joy in holidays. It's changed everything I thought myself to be. It's changed the world. It's a harder, darker world, a more frightening place than I ever thought it could be."

In addition to the crimes for which he's been convicted—the murders of Charlotte Murray Pace and Geralyn DeSoto—Lee has also been indicted on a first-degree murder charge in the death of Trineisha Dené Colomb of Lafayette and booked on first-degree murder in the deaths of Gina Wilson Green, Pam Kinamore, Carrie Lynn Yoder, and Randi Mebruer. He is booked but has not been tried for those crimes. The East Baton Rouge Parish District Attorney said he intends to seek three more indictments if necessary.

Lee now resides in a cell on death row in the Louisiana State Prison in Angola, under the care of Warden Burl Cane. That's in Lee's home parish of West Feliciana, but I doubt they are doing anything there to make him feel at home. It's said that one of the worst things about prison for the criminal is that they have to live with people who are of the same sort as themselves.

In December, before the death penalty was announced, Charlotte Murray Pace's sister, Sam, looked straight into the eyes of the shackled thirty-six-year-old Lee, in an orange prison jumpsuit in court. She'd be there to watch, she said, as "the first cold drop of saline hits your veins." *Good for her!*

JUSTICE: TOO LITTLE, TOO LATE

In February 2004, Zachary detectives said that DNA from a specimen found on a trash-can liner found at Randi Mebruer's home matched Lee's. It had been tested in 1998 only for fingerprints and blood types. Another first-degree murder charge was added to Lee's record.

WHAT EVER HAPPENED TO . . .

Chief Pat Englade retired on October 15, 2004, during that year's mayoral campaign. He cited job stress as one of the reasons for his decision.

Bobby Simpson lost his bid for reelection as mayor of Baton

Rouge. He was the first incumbent candidate in decades not to be reelected to a second term. He was beaten by Kip Holden, the first black person ever to be elected to that post.

Captain Paul Maranto, who was the East Baton Rouge Sheriff's representative on the task force, has been fired after being arrested for unauthorized entry of an inhabited dwelling, stalking, and hit-and-run. He has been ordered by the judge to seek help. I wish him well.

Sergeant Ike Vavasseur, who served on the task force and was one of the detectives involved in the initial search for links to the prostitute murders, is now commander of the Baton Rouge Police Department's Homicide Division.

Lieutenant Keith Bates is now chief of staff for Baton Rouge PD. He was one of the detectives involved in the initial search for links to the prostitute murders.

Mike Foster served out his eight years as governor. We don't hear much about him lately, but when we do, it's not very flattering. I take all that with a grain of salt. After all, this is Louisiana and politics are politics.

Mike Wolf, of WJBO radio, who vehemently condemned the actions of Lynne Marino during her quest to find justice for her child, is no longer with WJBO.

Jerry Fowler, now a widower, is out of prison.

Randi Mebruer's son lives with his father, Michael Mebruer. The ten-year-old knows that Derrick Todd Lee is the man who took his momma away forever. The continuing search for her body has been unsuccessful to date.

Lee Stanton, who was Carrie Lynn Yoder's boyfriend, graduated from LSU. The ceremony was attended by her parents.

Lynne Marino is still a victim's advocate. God bless her.

Ann Pace is still one of the most courageous women I've ever met. She remains an inspiration.

ANOTHER SERIAL KILLER

Baton Rouge, as we had suspected, was plagued by more than one monster. I wasn't there for this killer's capture, but I was on the receiving end of many of his victims. In April 2004, as Lee was awaiting trial, a SWAT team arrested a second serial murderer, Sean Vincent Gillis, a forty-one-year-old white male, at his home in Baton Rouge, for killings he committed over a period of ten years. Investigators said DNA evidence linked him to the killings of Katherine Hall in January 1999, Johnnie Mae Williams in October 2003, and Donna Bennett Johnston in February 2004. He was booked into Parish Prison on three counts of first-degree murder and three counts of ritualistic acts, including mutilating their bodies after death. He is currently accused of or has confessed to killing eight women:

- Ann Bryant, an eighty-two-year-old white female who lived in St. James Retirement Center. Murdered in March 1994.
- Katherine Hall, a thirty-year-old black woman with a high-risk lifestyle. Murdered January 1999.
- Hardee Schmidt, a fifty-two-year-old white female who was abducted while jogging near LSU. She was

similar in profile to Lee's victims. Murdered May 1999.

- Joyce Williams, a thirty-five-year-old black female with a high-risk lifestyle. Murdered January 2000.
- Lilian Robinson, a fifty-two-year-old black female with a high-risk lifestyle. Murdered February 2000.
- Marilyn Nevils, a thirty-eight-year-old white female from Abbeyville, Louisiana, with a high-risk lifestyle. Murdered November 2002.
- Johnnie Mae Williams, a forty-five-year-old black female with a high-risk lifestyle. Murdered October 2003.
- Donna Bennett Johnston, a forty-three-year-old white female with a high-risk lifestyle. Murdered February 2004.

Gillis was indicted on a count of first-degree murder and faces the death penalty for the strangulation of Donna Bennett Johnston. He was also indicted in Lafayette for the murder of Mary Nevils, was linked by DNA to three of the victims, and has confessed to the other killings. Some of the family members of the victims are not completely convinced that he is the killer and have adopted a wait-and-see attitude.

To say that Gillis or any serial killer is a strange duck is always a gross understatement. But there are some things that stand out in this case. For one thing, a geographical profile for where Derrick Todd Lee would have been residing was off by thirty miles or so; but that same profile was right on in Gillis's case. Ironically, it was based on some of Lee's victims. And while Gillis seemed to have a marked preference for stalking black victims in areas far from where he himself lived, the murder that led the task force to him, and to his arrest, was that of Donna Johnston, who was white, and he dumped her body by a bridge on Ben Hur Road, which is close to his home. She was found naked and face down. He left a tire track at the scene that was unique; there were only ninety sales of the Goodyear Aquatred 3 tire in the Louisiana area, and it was only manufactured for three years. Gillis's name came up on the sales list. Police went to his home, not far from the dump site of the latest victim, and swabbed him. They staked out his home until the rush job on the DNA was completed. It linked him to the murders.

Gillis's rap sheet was most unimpressive, yet his collection of reading materials included *Silence of the Lambs, Son of Sam,* and *The Hillside Strangler.* He had also been clipping news reports of some of Derrick Lee's homicides.

In an interview with Josh Noel of the *Advocate,* Gillis said that he felt the urge to kill only at night, and that in the four years when he worked the graveyard shift he never killed anyone.

Gillis was calm about the whole arrest. He elected to spend his last evening of freedom at home, awaiting the inevitable knock at the door. And when asked why he confessed, his simple answer was: "Because I did it." Not so simple was the chilling statement he made about the victims: "They were already dead to me."

The statement didn't surprise me. For one thing, these victims became objects and were devalued once under his control. It's always a hip shot in trying to understand what a serial killer means, for lots of reasons. On a gut level, a normal human is repulsed by the way these murders devalue life. One explanation for the statement is that if they were dead to him already, there was nothing wrong with killing them.

On Ben Hur Road there is a fading memorial to Donna Bennett Johnston. It's a heartbreaking collection of beads and wreaths. The saddest part is the weathered sign that just says MOM.

AND ANOTHER

What are the odds of having four serial killers in Baton Rouge, Louisiana, over the same time span? I have no idea, but we evidently have a fourth one. He's here now, still out there, still unknown, and still killing as of July 30, 2004, the date of his latest attack (at the time of this writing). He's been unofficially labeled the Black Prostitute Killer. His mode of operation and victim selection fit seven or eight murders, beginning in 1999.

The community seems pretty lackadaisical about the whole thing. Maybe it's battle fatigue. Maybe it's because of the victims' profile. I don't know. What I do know is that if we as a community do not learn from the past, we set ourselves up to repeat it.

FIFTEEN

Conclusion

"SO, WHAT HAVE YOU LEARNED, LOU?"

I guess the corollary of the Tin Man's question in the *Wizard of Oz* for me is: "If Baton Rouge is such a bad place, why do you stay?"

There are lots of reasons, family roots being a big one. Dorothy was right, there's no place like home, and this *is* home. I don't think Baton Rouge is so bad that it can't be fixed. Actually, I think that fix is happening now. We have a new mayor—Kip Holden. He's a good guy and a good leader. I recently saw him go to the crime scene of an especially tragic shooting in which three police officers were shot—one fatally. He was there at a time of crisis. He's involved. I respect that. I think we all do. We also have a new police chief by the name of Jeff LeDuff. He has presence and is the kind of leader who is able to gather support not only from his officers but also from the community. Incidentally, I think he is the first African-American police chief we have ever had. I hear the police department's morale has improved. So changes are in the wind. My bottom line: I have faith in the people of this parish.

As I glance back over these pages, the crime scenes jump out at me in vivid detail. I don't see cases or case numbers. I see an elderly grandfather who is a World War II hero going out into his backyard and putting a bullet through his head because he doesn't want to be a burden to his family. I see a small Vietnamese woman huddled into the corner of the convenience store, holding up her hands, trying to block the hail of bullets from the gun of the armed robber who kills her for no good reason. I see the body parts of someone's grown son, scattered for a half-mile along the railroad tracks after he was hit by an express train en route to the Super Bowl in New Orleans. I see the faces of six children who died in a fire, their little faces staring back at me as I remove the burned rubble from their little bodies.

Sometimes I wonder if maybe I have seen too much. There are hundreds of bodies of sons and daughters, husbands and wives, aunts and uncles, mothers and fathers, grandmothers and grandfathers, filed away forever in my brain.

Some of these people placed themselves in harm's way. Some others I don't have an explanation for . . . and never will. In practically every area of town I drive through, I can pick out spots where I have examined bodies. For a while, it was like some ghastly tour in the realm of death that I chose not to share with anyone. The places and the memories are still there, but they have no power over me any longer. Usually they just sadden me. My nightly bicycle ride takes me past the office of a colleague who killed himself there several years ago. He was a good man. As I pass by, I always say a prayer for him. Maybe the prayer is for me.

I have seen the casualties of those who lost loved ones, scars that never heal and leave them asking the unanswerable—why? I still have people stop me at the mall and on the street to thank me for what I did and to give some follow-up about themselves and their families. Baton Rouge truly is an oversized small town. Not everyone is satisfied with the job I did, but I did it to the best of my ability. I tried to be a true public servant; that may sound hokey, but it's true.

I've seen people at their worst and people at their best. A moment of rage due to betrayal (real or imagined), a drink too many, an extra hit of Ecstasy, a cocaine run, a fit of jealousy, passion, a moment of depression and hopelessness—any of these negative emotional experiences can result in homicide, suicide, or accidental death. I sort of understand. I don't condone the behaviors, but I understand the cause and effect.

What I do not understand is the sociopath who kills for the fun of it . . . because he likes it. I don't understand the pedophile. I don't understand the child-murderer. I don't understand the serial killers and rapists. I just know they are evil. I have sensed that evil still lingering about the death scene. I'm neither psychic nor psychotic, but there is no other explanation for what I have witnessed on such occasions.

I've also seen the toll this takes on responders. Pick one—EMS, fire, police, coroner. Look close and you'll see the scars. Some may be just superficially sealed over—like a scab trying to cover an infected wound. It's an expensive cover-up. Callousness, failed marriages, depression, chemical dependency, alienation of family, cynicism as a permanent character trait, identification with the perpetrator and the subsequent attitude that the end justifies the means, suicide . . . yes, even homicide. I've felt the rigid, steel-clad silence at a murder scene as I strained to pull the dead lover of a policeman's ex-wife out of the trunk of his car. There were no police lights flashing in the darkness because the killer was one of their own. Intense . . . It's all there, just look a little closer sometime and you may see it. It's also avoidable, to a large degree. And I'll say it again: If you don't process all the pain, it will process you. Anyone who doesn't believe that has missed one of the lessons I have tried to offer in this journal. I don't want any responder to end up a corpse in a bathtub.

I've seen Buddy and Betty Knox, Jim and Edith Moore, and others who were among the founders of the LOSS group, get out of bed at all hours of the night to head to a crime scene, responding without question to four words from me: "I've got a suicide." I've seen them share their pain and their hope as they help others navigate through a disaster.

What do you say to a grieving wife and mother who finds

her husband dead in his bedroom, having killed himself with a shotgun blast to his head? Or it's 2:30 P.M. and the kids are in school, and a horror awaits them at home. Who is there to prepare the way for them? People like those of the LOSS team. They stay the night in a bloody trailer with a grieving wife who has no family or close friends in the area to stay with her. And never once did I ever hear any hesitancy in their voices. As an aside, I'll just mention that Jim Moore is seventy years old, but I've never seen him flinch from his self-appointed duty.

I've seen a town come together over the body of an abandoned child. Folks still put toys on her little grave. There was a play cell phone there on her headstone just yesterday when I went by to visit. Christine Noel Love has been adopted and not forgotten.

I'm reminded of memorials that family members have erected to their lost loved one. Most of the memorials are temporary and in time melt back into the landscape. Still, you don't have to go far to find a cross on the side of the road. While sometimes fleeting, some of those memorials really stick out in my mind. The teddy bear and toys that marked the tragic death on College Drive a half-mile from my home are particularly haunting. It's where a baby of less than one year of age was catapulted from her stroller when a drunk driver hit her. The driver sped away with the stroller still lodged under the car's bumper. It was December 19, 2002—Christmastime— baby's first Christmas. Her Christmas gifts were placed at the site of her death as a memorial.

The media has hurt—and at times helped. Greg Meriweather, who does the street scene for Channel 9, is a straight shooter who is really interested in making a difference, not in merely sensationalizing. He's done some neat things, like trying to get people to keep their kids away from homicide crime scenes. When we had four infants die as a result of rollover in a parent's bed, Phil Ranier, a TV health reporter, came forward. He interviewed one mother, and the story made such an impact that we went many months without another rollover fatality. It's not all about death.

I've worked alongside a lot of cops, both uniformed and detectives. The really good ones approach each crime with the same intensity and dedication. It doesn't matter to them if the victim is a "throwaway." Each person gets the same attention. Those are the great ones. They seem to rise above the political and departmental pressures of the moment to do their job. They don't talk about it much, and they don't go around patting themselves on the back. This is serious business and they are serious men and women.

I have struggled for years with that little saying "Let go and let God." It actually used to irritate—then frustrate—me. Truth is, it pissed me off. I never was very good at letting go. I like order, and since this is the time for true confessions, I must admit that I'm something of a control freak.

I've tried to figure it out and answer the "Why?" question. With all my philosophy, religious study, and intellectualization, I have only been able to come up with one answer: "Because." There must be an explanation, but it transcends all the reasoning power that I have been able to muster. In the end I have had to let it go and give it over to God. No, I don't think of God as my "fall guy," but as the ultimate source of reason and wisdom.

"Let go and let God."

I hope I can keep it there.

ACKNOWLEDGMENTS

It was during the worst times, in the midst of the troughs of the serial killer's reign, when I met Patricia Cornwell. She came to Baton Rogue offering help and was generally rebuked by the serial killer task force and local government. I remember the mayor telling me that she was just here to write a book. He was as wrong about that as he was about his chances for reelection. She came to help and she brought expertise and experts with her. They, too, were welcomed like seeds upon the rocks.

I can't say I was much better initially. There was a media feeding frenzy going on. My policy to date had been to deny major media interviews and keep this at a local level. I didn't want to empower the killer any more than was already done. I just wanted to get the facts out. And we had some really sincere reporters, like Greg Meriwether and Avery Davidson, whom I have found to be most responsible in their reporting. In short, I didn't need those national guys to dispel rumors and get the right messages across.

Caron Whitesides, my executive assistant for many years, came into my office one day and announced that Patricia Cornwell was there to see me. And while I knew who Patricia Cornwell was, I didn't really know who Patricia Cornwell was. Most of my "set-aside reading time" is spent on forensic and mental health books and journals. But Caron seemed to know and respect her and so did my wife. Caron has always covered my back, and her opinions weigh in heavily when I make such decisions. So, though I was somewhat irritated about the way they were hammering on me to see Patsy, I acquiesced, mostly just to get them off my back, and said I would meet with the lady. "But just for a minute," I admonished. It was one of the best decisions I ever made. It took me no time at all to realize this woman was the real thing and, more than that, she cared about my people. I came to know and trust her, and down the line I shared my journal with her. She encouraged me to move forward with it.

She, in turn, put me in touch with Kris Dahl. My God, I had an agent! Kris is great, and she led me through the morass of publishing. I must say folks at Putnam have been most patient with my amateurish and naïve approach to this endeavor. Kris and Putnam introduced me to a fellow Louisianan, albeit one who had become a galvanized New Yorker. His name is Mark Robichaux. He helped me put things into an organized format and was officially my "collaborator." He's a lot more than that—he's also a friend. We got to know each other when he came down to Baton Rouge and got the tour of various sites at which some of these homicides had occurred—definitely not the kind of tour most people get.

During my tenure as coroner, I have had some great teachers. Mary Manheim shared openly with me out in the field and in her anthropology lab at LSU. Lamar Meek was not only an excellent forensic entomology professor, he was also an inspiration. Neither one of these underpaid consultants ever said no when I needed them. And then there is Dr. Michael Cramer, one of the most conscientious forensic pathologists I have ever seen. He excelled at one-to-one teaching in the morgue. I've been through many an autopsy with him. If it was there, Mike would find it.

I am eternally grateful to Buddy and his wife, Betty, who lost their son to suicide, and to James, who also lost his son to suicide, and his lovely wife, Edith. These folks were part of the founders of the LOSS program. Each would get out of bed at all hours of the night and head for whatever part of the parish I was in. They would do so without hesitation when they heard those four chilling words from me over the phone: "I've got a suicide."

Wanda Hebert served as the moral compass of my office and taught me the true meaning of the term "iron hand in a velvet glove" by the caring but firm way in which she dealt with our psychiatric patients.

I have also learned a lot from the numerous detectives I have worked with. I learned what to do from the good ones and what not to do from the not-so-good ones. I owe a special thanks to the crime-scene investigators, who always took the time to teach me.

And when it was all said and done, DeAnn helped me get past my reluctance to expose my thoughts and feelings about being a coroner. And so, here we are. We have a book and I have lots of people to thank and to be thankful for.